THE
INVISIBLE HAND

The Triumphs and Sacrifices of Military Service

SGM PAMELA DUGGAN (RET.)

Copyright © 2026 by Pamela Duggan

Publishing all rights reserved worldwide.

All rights reserved as sole property of the author.

The author guarantees all content is original and does not infringe upon the legal rights of any other person or work.

No part of this book may be reproduced, stored in a retrieval system, or transmitted in any form or by any means, without expressed written permission of the author.

Edited by Lil Barcaski

Published by: GWN Publishing
www.GWNPublishing.com

Cover Design: Kristina Conatser and Pamela Duggan

Developmental Editor: Quinton Marcelles

Artwork by Stoney Trey Anderson

ISBN: 978-1-965971-38-3

BOOK MANTRA

*"Behold it came quickly. A life of sacrifice.
He that have an ear, let him hear."*

DEDICATION

I dedicate this book to my loving parents, Bennie and Lucille W. Johnson, who raised me to love and know God, to be accountable, confident, and responsible.

All my love, appreciation, and gratitude to my husband Calvin, and my two sons Bryant and Stoney, who stood strong with me as I advanced through my military career.

TABLE OF CONTENTS

Book Mantra ... 3

Dedication ... 5

Endorsements ... 9

Introduction ... 11

CHAPTER 1: In The Beginning ... 13

CHAPTER 2: My Steps Are Ordered ... 33

CHAPTER 3: Challenges, Changes, and Choices (Three Cs in life) 45

CHAPTER 4: The Sound of My Yes ... 67

CHAPTER 5: From Orders to Outcome ... 87

CHAPTER 6: What You See Is Not Always What You Get ... 117

CHAPTER 7: The Pivot Space ... 123

CHAPTER 8: To Every Purpose There is a Time and Judgment . 131

CHAPTER 9: Making a Difference ... 137

CHAPTER 10: For the Time is at Hand ... 151

CHAPTER 11: Retired Not Expired ... 157

CHAPTER 12: Moving Forward, I Am Alive Forever More ... 165

A Final Note ... 171

Acknowledgments ... 173

About the Author ... 177

ENDORSEMENTS

I have experienced first-hand the mindfulness of SGM Pamela Duggan's professionalism and the impact of her actions through leadership. She holds herself first and then everyone involved to a standard that fosters success. Pamela's thoughtful humanitarianism gives birth to actions that lifts others from hopelessness to believing in life again, and when it mattered most, she was there. Additionally, Pamela is able to anticipate the needs of others through the strong sense of intuition she possesses. She moves in accordance with our NCO creed by exhibiting a high degree of skill and mastery for making others better.

Leon Mangum, MBA

Pamela Duggan distinguished herself early as a tenacious, mission-focused Soldier who lived the Army Values without compromise. Long before attaining the rank of Sergeant Major, she led with the presence, competence, and character of a senior leader. She never waited to be asked, never avoided difficult tasks, and always gave 100 percent.

Known for her integrity, professionalism, and unwavering commitment to Soldiers and family, Pam inspired trust and lifted everyone around her. Now retired and an author, she continues to serve through her words and wisdom. This book is a powerful extension of her lifelong calling to lead and inspire.

MAJ Sheila Watford, U.S. Army (Ret.)
& SGM Milton Watford, U.S. Army (Ret.)

Congratulations to SGM(R) Pamela Duggan, one of my most esteemed mentors and leaders from my military career, on completing your first book! You are a pillar of professionalism and possess a natural ability to lead. I have preordered your book and look forward to adding it to my collection. Your impact on my life and the lives of soldiers like me has been profound. I wish you continued success and blessings.

MSG Desiree Pabon, U.S. Army

Pamela was an instrumental leader to me by showing that through the valleys of hardship one can get to the peak.

MSG Jamie Jacobs, U.S. Army (Ret.)

The inspirational life of an inspiring warrior and lady, SGM Pamela Duggan's story is one of determination, resilience, and faith in the face of events that would break many.

Her life and success in environments that were not well suited for her will motivate you to keep going and keep striving.

LTC Brian Jones, U.S. Army (Ret.)

This book is a powerful reflection of a life guided by faith in God and shaped by service, discipline, and perseverance. Pamela's story encourages readers of all ages to remain anchored in faith while pursuing their God-given purpose!

Rory & Kahlila Lawrence

INTRODUCTION

Life: a period between birth and death. A journey filled with functional activity, and continual change, all unfolding in a series of single moments. It is a matter of choices, both good and bad, that define us. It is important to me that I leave a memoir of my personal experiences as a daughter, mother, grandmother, Soldier, mentor, and leader for my sons, family, and friends. Through this book, I hope to help them understand "The Why," how, and will, that prepared, and motivated me, to serve this great nation for 37 years of my life.

Do not be conformed to this World, but be transformed by the renewing of your mind, that by testing you, you may discern what is the will of God. (Romans 12:2)

My faith has provided me with complete trust and confidence in the people and systems that I have lived within and navigated throughout my journey. As the reader, I invite you to take a tour through my life's choices and challenges, and witness how the decisions I made and the invisible hand that guided me have shaped my path.

The Ordinary World

We meet our hero.

CHAPTER 1

IN THE BEGINNING

"Perhaps you were born for such a time as this."
Esther 4:14 (KJV)

My parents' love story began in the innocence of youth, crossing paths with destiny and faith long before I ever came into the picture. My father was born on April 28, 1923, in Jackson, Mississippi, the only child of Daisy, a devoted housewife, and Albert Johnson, a gifted chef who owned his own bakery. My father's upbringing was quiet and steady, surrounded by the values of hard work and pride in one's craft. My mother, aka "Sweetie" was born February 15, 1930, east of the Mississippi River in a modest town called East St. Louis. For all my life, my mother told me her birthday was February 14, Valentines Day, and it wasn't until her golden years I learned she was really born on February 15. Her mother married young and had four children but lost her husband to a critical illness. She remarried, and my mom was the first of two children born to Charles and Emma Young Fields Walters.

My mother often shared the story that she was the only child born from the union of Emma and Charles Walters, and that her mother died giving birth to her. I learned a different story from my older cousins after my mom passed. I learned that her mother gave birth to a second child that passed away before the age of one and then shortly after passed away herself. My mom was almost one year old when she died so she never knew her mother or that she had another sibling.

It is funny how the stories of yesterday never seem to be told until centuries later, and by then the truth is only what they can remember. My mother's life started with heartbreak. After her mom passed, the four children from my grandmother Emma's first marriage went to live with their father's family, and my mom grew up apart from her brothers and sisters. My mother's heritage is rooted in a mix of Native Indian, and Irish ancestry. East St Louis, Illinois had a long inhabited Native American culture on both sides of the Mississippi River. Her father was one of four children. He and three sisters were fifth-generation descendant of the great Geronimo, the revered Chiricahua Apache tribal leader, medicine man, and war strategist who fought fiercely to protect his people against U.S. and Mexican forces. Grandfather Walters was known as "Big Chief" in the city. He was a proud man who owned a grocery store and carried the wisdom of his lineage. He told my mother she was born with a veil over her eyes, a sign believed to mark those with spiritual gifts, what some call a sixth sense.

My mom was raised with wisdom beyond her understanding, and all her life she knew there was a presence with her. She would tell us stories of how her ancestors would visit her when she was growing up. Her father told her to never be afraid but always ask them "what do you want?" She followed what her father told her and every day of her life, her spiritual gift guided and spoke to her, oddly enough, only in her right ear. Her childhood, however, was filled with loss. After being raised in East St Louis, Illinois by her father for 14 years, he fell ill and passed away. To my understanding Mom was raised in her family home. After her father passed, her aunties took on the responsibility of taking care of her.

At sixteen, instead of celebrating her "Sweet 16," my mother faced another heartbreak: her Auntie and first guardian was tragically killed, forcing her to leave everything familiar behind. She moved

In The Beginning

to Chicago to live with her Aunt Catherine, where she found stability long enough to finish high school. After graduating, she returned to East St. Louis, where her Aunt Emily, a coal company owner, became her legal guardian and a steady force in her life. Even in her adult years, my mother cared for Aunt Emily until her passing. As a little girl, I often went with her, watching as she gave medical care, did the shopping, the laundry, and kept the house in order. Aunt Emily had simple joys: an ice-cold Coca-Cola, sugar-dusted lemon drops, and the company of her beloved cats.

My mother found joy in taking care of others, music, dancing, and playing card games like Spades and Bid Whist, a game that goes as far back as the 1800's, created by black folks in the south. One weekend, while out enjoying herself at an upscale local lounge called Pudgey's, she met my father. He had moved from Jackson, Mississippi to work for Chessy, B&O Railroad. She was 17, young, beautiful and built like a Coca Cola bottle. Her skin was Carmel brown, she had high cheek bones, and her hair was jet black and so long she could sit on it. She wore tailored clothes and high heels. My father said she looked like a China doll. She was sassy, classy, and full of life. His skin tone was high yellow; they called him "Red." He liked dressing in top vests, three-piece suits and Stacey Adams shoes. He wore a pocket watch and a top hat; he was 23 and knew he had it going on.

They dated briefly and fell in love quickly. They married just shy of my mom's 18th birthday, but Aunt Emily had the marriage annulled because of my mother's age, and told her she was too young, and he was too advanced for her. Still, love finds a way. As soon as she turned 19, they remarried. We used to laugh, and say Dad thought Mom was so nice, he married her twice. And from that day forward, they stayed united loving, laughing, enduring through 52 years of marriage, until my father passed away Easter Sunday morning, holding mom's hand, on March 27, 2005, at the age of 82.

THE INVISIBLE HAND

Dad and Mom.

I never had the opportunity to meet my grandparents. They had all passed before I was born, yet their legacy is stitched into the fabric of my being. After she married, my mother set her sights on becoming a nurse and enrolled in nursing school. During the summer, she lovingly cared for her young nephew Gregory so her sister Lillian could continue working. Balancing textbooks and toddler toys, she pressed forward with determination, holding tight to her dream. But in her third year, life shifted. She discovered she was expecting her first child. Not long after her first trimester of pregnancy, my father asked my mother to set aside school and become a full-time mother. With quiet strength, she laid down her dream of becoming a nurse and poured her heart into raising their family. My father, an only child, always longed for the joy of a big family. He dreamed of nine children filling their home with love and laughter. My mother, devoted and willing, tried to honor that dream. But after giving birth to four children, her health began to falter. She developed kidney failure, and doctors presented her with a devastating choice: take the medicine that could save her life but harm the baby she carried or refuse treatment and risk losing them both. In that moment, my mother chose faith. She said the Holy Spirit whispered to her, "Always trust God. Only God can give life and knows what tomorrow holds." With unshakable conviction, she told my father that everything was going to be alright and that she would carry the pregnancy to term.

My father had witnessed before how the Holy Spirit revealed truths to her that later came to pass, so he trusted her completely. Together, they leaned into faith over fear. The pregnancy was carried through, but it came at a cost—my mother lost one of her kidneys in the process. After giving birth, she began treatment, and miraculously, she went on to live with just one kidney for more than sixty-five years. In 1966, doctors told her she would only survive another ten years. But God had other plans. She lived to the age of ninety-four, never once needed dialysis, and never lost her second kidney. Even in her nineties, though she faced challenges like incontinence, she continued managing life with only one kidney. Her life stood as a testimony not just of resilience, but of faith that defied every medical prediction.

Their love story was never just about romance. It was about sacrifice, commitment, and belief in something greater than themselves. It was about two people rising from loss and both without siblings growing up, leaning on faith, and choosing each other again, and again in youth and age, in sickness and health, in laughter and tears. Their love shaped my understanding of life and purpose, and it continues to guide me today.

However, no sacrifice comes without cost. My father had to work tirelessly for 33 years, never taking one day of sick leave, holding two jobs, one as a railroad worker, later becoming a railroad engineer for B&O Railroad Company, and then as a welder at the steel mill in St. Louis, Missouri to provide for his family.

Mom loved very nice things, and she provided nothing but the best of everything for her family. Her favorite store was Neiman Marcus. I remember the story about my siblings being the only children in the neighborhood that had a play gym in their back yard. Since my dad worked on the railroad, she could travel anywhere free on a railroad pass. She said it cost 50 cents to cross the Popular Street bridge into St. Louis, Missouri, but she could ride the train anywhere the railways traveled for free. She would take the train to New York to buy our clothes, because she never wanted us to look like everyone else. Once, she told my dad that she wanted my younger sister and I to have 12 new dresses to start the school year. My father said he could not afford to get 12 for each of us. My mom said "OK", then she went downtown in East St Louis to this boutique dress shop for children, and got a job to buy us those dresses. When the school year started, we had 12 dresses, four pairs of shoes, and two coats each. I was in the 5th grade that year, and Ms. Tony, my teacher, asked, "When are you going to stop wearing new dresses?" because most families buy new clothes for the first day back at school, but my mom believed in being unique. Mom did not like us sharing clothes, so she always made sure each of her children had their own stuff.

She was a faithful member of our school district 189 board of education, the Parent Teacher Association, and would attend the school education meeting in the town city hall. She was an advocate for our community, always in government meetings talking with the town leadership, fighting for better opportunities in school. She would attend regular events at the local Library, she met President Barack Obama in the East Saint Louis library in 1997 when he held a seat in the Illinois State senate, She was an Eastern Star, and belonged to so many community leadership panels.

I was born in St. Louis, Missouri, at Saint Mary's Hospital, just across the Mississippi that divided the state line between Illinois and Missouri. I grew up in the 60's, in East St. Louis, Illinois. Now, I know some people have heard that East St. Louis is the worst city

In The Beginning

to live in, nothing but deteriorating homes, crime, and poverty. There may be some truth to that today, but East St. Louis has a rich history. In 1958, East St. Louis was named an "All American City" by the National Civic League, having developed an industrial core with railroad-related industries and warehouses and the population was over 82,000 residents. In the 1950s and beyond, East St. Louis hummed with music. From smoky blues clubs to lively street corners, the city's musicians poured their hearts into blues, rock and roll, and jazz, shaping sounds that would ripple across the nation. Some famous stars like Ike & Tina Turner left to claim the spotlight on a national stage, while jazz great Miles Davis, born in nearby city called Alton but raised in East St. Louis, carried the city's rhythm and soul to the world.

When I was growing up, there were many businesses thriving, with employment opportunities, good public and private schools, and colleges with many degree program choices. I was one of five children, two brothers, Michael and Cecil, two sisters Stephanie and the youngest sister, Taceillya and me, the middle girl. I am sure you have heard about the cliche about middle children being different, that we are special and unique. Born in August under the Zodiac sign of Leo, I have always been a natural leader who shines in social settings, always willing to take risks and stand up for myself and others. Constantly striving for success and recognition, often with big dreams, perpetually drawn to art, drama, and self-expression, I possess many of the characteristics of a Leo that is ruled by the Sun, which gives me a bold, radiant, and magnetic presence. My mother passed down rich cultural and spiritual traditions like prayer before eating and going to bed. Every holiday was a special treat. At Christmas, we always had plenty of gifts, but we could only open one gift the night before Christmas day, and every year it would be a pair of new pajamas. I have kept that tradition with my children and now my grandson, along with other traditions like many family recipes such as egg custard pie, how to set a proper table with fine China and gold silverware, preparing black eye peas with money, mixing the good luck coleslaw, and homemade biscuits, steak and rice on New Years Day, baking the Easter cake with three layers, yellow, green, and pink with chocolate icing.

From a young age, my mother predicted I would be a communicator. At just one and a half years old, I would wake up at 4:00 AM to have coffee with my father before he went to work. Though he never understood what I was saying, I always had something to tell him. My mother called it "blubber" or "juba," but to me, it was communication. We lived in a high-rise building called "Orr Weathers." There were four buildings labeled B, C, D, E eight stories tall. It was considered a very nice area. There were sidewalks all through the community with beautiful grass chained off so no one could walk on the grass. It had two large playgrounds, one for younger children with swings and sliding boards, see saws, and sand boxes and an area for older children with monkey bars and taller slides, and swings with long chains to go very high. The front of each building had a porch and a seating area outside the front, and every evening in the summer all the families would join there. The children would play until the streetlights began to slowly come on, then we knew we had better make it to the front of the building to meet our parents. It was a very nice area that offered huge apartment homes and residents were required to present a marriage license to live there.

Our home was building E, third floor, apartment 301, a corner unit with a living room, dining room, kitchen, and three bedrooms, one for my parents, one for my two brothers, and one for me and my sisters. It was spacious. I remember our bedroom had a large window and at night I could see the St. Louis, Missouri skyline and the famous Gateway Arch. But there was only one bathroom; my mother had to set a strict schedule for us to use it before school and bedtime. They were perfect role model parents. My mom told me she did the best she could, and she hoped she was a good mom to us, and I never missed an opportunity to tell her that she was a great mom, and I loved her.

In The Beginning

We all went to the appropriate grade level schools in the area. My elementary school was Martin Van Lucas, mascot a tiger, grades first through sixth. By the time I entered middle school, my mom and dad had purchased a three-bedroom home off Lake Drive on the east side of town. My middle school was Clark Jr. High, mascot a red cardinal, grades seventh through ninth. By the time I entered high school, my mom and dad purchased a larger home from a family that had two daughters training for the swimming competition in the Olympics. The family had a full swimming house built in the backyard of the home with an indoor swimming pool for lap swimming. It had a complete bathroom, large glass sunlight, wood deck, and a full entertainment area. My high school was East Side High, the Flyers (an airplane mascot).

East St. Louis is known as the city of champions. We earned that name because the middle and high schools have several state championships within our sports programs. The middle and high schools are very successful with creating athletes in track & field, football, and basketball. We have had several successful athletes and Olympic Medal winners come from East St. Louis, like Al and Jackie Joyner. They are the best-known celebrities in my city.

I wanted to write this book because I want to help some of those people who read it make better choices. I want to share some things about life that often are not shown to you while growing up. I know we all have stories; some we want to share, and others that we don't. I really think that it is important that we speak and say something about the things we've learned because it may help someone else make a better choice or use our example as a road map to become better, or greater. Everybody is different, and we all have something that we feel is most important to us, but what I've learned as I have grown wiser, is that it is equally important to share what that thing is. Because someone may get some help from that thing you didn't feel was important.

THE INVISIBLE HAND

As children we are all influenced by something whether it be traditions, culture, ancestors, or circumstances. I remember sitting at the kitchen table as a child, watching my mom juggle a hundred things at once, preparing dinner, helping with homework, and still finding the energy to listen to every worry I had. She moved through life with a composed determination and a confidence that seemed effortless, yet every action carried purpose. It was in those moments that I first learned what it meant to be resilient, adaptable, and determined, qualities she embodied daily and unknowingly passed on to me.

Taceillya, Stephanie and me.

Growing up in the 1970s, I also looked to women outside my family who inspired me. Pam Greer, the actor, activist, and strong

African American icon, showed me how courage and intellect could challenge the world. And then there was Wonder Woman, the superhero who fought for justice and equality, representing independence and moral conviction in a way that captivated my imagination. In many ways, I see my mom and these role models reflected in the woman I have become. Like them, I strive to combine strength, intelligence, endurance, and compassion, standing up for what I believe is right while inspiring others to do the same. Their influence, intertwined with my mom's teachings, has shaped the way I navigate life with courage, adaptability, and a steadfast commitment to making a difference.

My brother Michael.

My brother Cecil.

My mother was a great mom, one of the best moms any child could have, and she was proud of her life and life choices. In the same spirit of faith and sacrifice, my parents opened their hearts and their home by adopting their grandson (my nephew) Vergil, determined to give him the stability and future every child deserves. My mom and dad were intentional in every detail of his upbringing ensuring he was surrounded by people who modeled professionalism, integrity, and purpose. Vergil excelled academically, standing out in chess, math, and science, and in time, he chose to follow a path of service, answering the same call to serve our nation by joining the United States Army. He went on to serve in the Special Forces, building a distinguished career that culminated in his retirement after 24 years of honorable service.

Mom made many sacrifices for herself to ensure her children had the best of everything. She was an excellent cook and could make any dessert. She never used boxed products, everything we ate was homemade from scratch. She prepared homemade meals daily, desserts, potato chips, and donuts weekly. My father would bring home large containers of orange, and lemonade Tang powder drink mix, along with chocolate, and strawberry Nestle Quick mix, boxes of treats from the candy store, thick suckers shaped like an octagon that had a little bird raised on the top (the flavors were lemon, orange, grape, cherry, and green apple), a barrel of huge pickles, large long taffy strips mixed with banana strawberry, and blueberry flavors , and candy bars like Zero, Mr. Goodbar, Almond Joy, Payday, 100 Grand, Milky Way, and Nestle Crunch that my mother would add to our lunch bags. She often added little notes of encouragement to carry us through the day.

She was very present in our education, frequently visiting our schools and ensuring we were on our best behavior. Back then, parents were highly involved in schools, and my mother was well known throughout the school district. She also volunteered in the classroom and at government offices. She made sure we were engaged in political resolutions, congress legislation, and local and federal laws pertaining to our livelihood, education opportunities, and all our extracurricular activities. She was a classy and proper woman, always wanting the best for us. She bought us several Avon fragrances, and cologne to ensure we smelled fresh, combed our hair with fancy ribbons, and barrettes, and dressed us in clean white Keds tennis shoes, Buster Brown shoes, and pressed

In The Beginning

clothing. I remember her making me wash my shoelaces on an old scrub board every night. My childhood was filled with structure, family, friends, fear of God, faith, and love.

I started preschool (then called Head Start) at the age of four, walking to school with my mother every morning. In the afternoons, she would pick me up and I would play on the neighborhood playground. Evenings at "Orr-Weathers," our home, were filled with outdoor games like Tag, Hopscotch, Hide-N-Go Seek, Little Sally Walker, Double Dutch, and marbles. Our friends often let me go first, perhaps because the chalk, ropes and marbles were mine, but it instilled in me a lifelong belief that to be first, you had to bring something to the table. Our family was active in the Seventh-Day Adventist Church, where we could not do anything from sundown Friday to Sundown Saturday. For entertainment, my brothers would pull me on a cotton blanket down the halls of our home. That kept the floors shining and one less thing for Mom and Dad to clean.

Me, at the age of 4.

Many of the members of our church owned businesses, including a beloved ice cream truck called Mr. Frosty. We attended many Church revivals and choir events. My older siblings and my mother were members of the choir, so that old saying about having something to do at the church all week was true. My dad made sure my mom had a vehicle to go where, and when she wanted. She would pack lunch, and us, in the car, and drive us to every event. We also enjoyed holiday celebrations, and traveling on the train first class to summer vacations to visit relatives in Chicago, Ohio, and once, a trip to Disneyland in California. These family activities enriched our childhood. My mother would take us out to dinner once every two weeks to teach us dining etiquette, the importance of ordering at restaurants, and how to interact with people respectfully. She also enrolled us in the YMCA where my siblings and I learned to swim. I learned to swim in the deep waters at the age of four. We were the first Black family featured on the YMCA's advertisement cover in Belleville, Illinois. Many people commented that we were raised "like white children," but to me, it was natural. We had bikes, roller skates, and skateboards. I always believed that all children had similar experiences to mine.

During the summer months, when I entered the seventh grade, I would set up the tables and chairs on our front porch and tutor children on math and reading in my neighborhood, setting up workbooks and snacks for those who had trouble learning, or who had missed school. Years later, one of the young men (my nephew now) credited me with inspiring his educational journey. He graduated from college, and his daughters attended college with full scholarships. In school, I was a very good student, and athletic. Starting with cheerleading in elementary school, by middle school I continued cheerleading and joined the dance team, track and field, and played the clarinet and drums in band. When I reached high school, I continued playing in the band, cheerleading, track and field, and started playing basketball, tennis, and excelling in gymnastics.

In The Beginning

Me around the age of 7

I was on track to attend college and had dreams of receiving a scholarship as a cheerleader, but during my senior year of high school, my mother took me off the cheerleading squad because of an incident that happened with a male head coach while I was away at a competition. I told my mother I was in my assigned room with my squad getting ready for the game, and the head coach entered our room and asked us what we were doing. My captain got up and said, "We are just getting our hair and cheers together." He stepped to me and asked my name and said, "I need you to come outside and talk to me." I asked my friend to come with me, but he said just me. I said, "No," then he replied, "Well, if you can't do what I ask you to do, you are off this cheerleading team."

The next morning, I told the head cheerleading coach what had happened and when I returned home, I told my mom. She went to that school and spoke with the principal. The principal had been my six-grade science teacher, so he knew me and my mom, and he knew I would not make something like that up. Actions were taken and I was offered my position back on the team, but my mom said, "No." That shattered any dream of mine being selected for a college squad. I had cheered on a team since the fifth grade; I was truly heartbroken. The head cheerleading coach made sure I received all my senior letters and pins at graduation.

My high school letters.

When I graduated high school, I had three options: get a job, go to college, or join the military. My siblings had all joined the Marines when they graduated high school, but at that time, I was not trying to be one of "The Few and Proud Marines." I was dating my high school boyfriend who was a typical 70's teenager that balanced freedom and rebellion. He owned a motorcycle and eventually taught me how to ride. He had family awkwardness, but he graduated in 1979 from high school to join the U.S. Army. When he returned from basic training, he proposed marriage, and we were

In The Beginning

married December 29, 1979. At the age of 17, before I graduated high school in May 1980, I became a military wife. Once I graduated from high school, I moved to Ft. Bliss, Texas, and was there for about six months when he received orders to report Germany. I had to stay behind and wait for command sponsorship to join him in Germany. While I waited, I returned home to my mom and dad. We stayed in contact while we were waiting for a housing unit to open.

Six months passed, and my husband stopped contacting me. I made calls to the number he gave me and wrote to the address I was provided, but I didn't get any responses. A few more months went by, and instead of hearing from him, I was contacted by the military police. I was asked about his whereabouts and if I had seen or heard from him. I answered the questions revealing that I had been in contact with him up until about two months prior and I had not heard or seen him. The Police informed me that he was under investigation for something to do with narcotics. Since the police were looking for him, I knew that he somehow evaded the authorities and intentionally avoided communicating with me. I never saw or heard from him again until he contacted my mother sometime later saying he had been discharged from the Army and had returned home. That was my first eye opening experience about trust. I trusted him with my future, and he failed to follow through, and now I had to live with the circumstances of his choices.

Clearly my dreams were destroyed, I had given up my opportunity to go to college and chose him over my own future. I had no money, no education, and now I had to stand on my own and make a living for myself. I was not informed of any resources or programs that were available to assist me as a dependent wife. I had to get a job and begin figuring out my future life. I started working at the local McDonald's Restaurant. I worked six days a week and, in the evenings and weekends, I would go to the skating rink. That was the local hang out and I loved skating. After about eight months, I got began a relationship with a friend from the skating rink.

After a few months of dating, I learned he was not being truthful about his life. While downstairs at his home, I overheard an argument between him and his auntie upstairs, which I learned then were not related. He was saying "Auntie, let me explain," and she replied "I don't want to hear another lie out of your mouth. I let you live here to help you out. Stop calling me Auntie; you are not my nephew. You don't have a job, and you haven't paid any bills here for months. I want you gone." He returned downstairs and I acted like I did not hear their conversation. He said, "Let's go, I need to get out of here." I asked where we were going, and he said to his sister's house. During the ride over, he began to share his side of the story, and he was going to stay with his sister until he got things sorted out. I went home, continued working, and I would visit with him. He lived in the basement of his sister's building, and often would not have food to eat. After a month, he thought it was a good idea for us to get a place together. Love is blind, but my mother didn't raise no fool. I still lived at home with my mom and dad. I refused and he became furious. It was late in the evening and I told him I was leaving; then I went home. When I arrived home, my dad was at work, and I told my mom what happened. My boyfriend decided to follow me home and confront me. He was beating on the door consistently, and my mom said, "If you don't leave, I am calling the police." He finally stopped and slammed the porch door and decided to remove the front gate from the hinges on his way out. He threw the gate in the trash can up the street.

My mom called my dad and my brothers. When they arrived at the house they told me to take them to where my boyfriend lived, but when we arrived, he was not there. My father said "Ok, we will catch him." We returned home and my dad and brothers fixed the gate. The next evening my father and brother accompanied me to the skating rink to catch up with him. When he pulled around the corner, I saw him looking to park his car. I told my dad, "There

In The Beginning

he is, parking his car." He locked the doors and started walking toward the skating rink entrance, but when he saw me, he took off running across the open field next to the building. My brother shot a gun in the air, over his head to scare him, but he kept running and soon disappeared into the wooded area. My brother drove around looking for him, but we never saw him again. My dad said, "We will get him if he comes around here again." At home we had a screened in porch with large roll out glass windows. Till the day my father got too sick to do so, he sat on that porch every day. He would sit in a chair near the screen door, prop his feet up on the wall that faced the street near the mailbox, and people-watch. Funny how things can shape your future. After that, I knew it was time to move on and do something else with my life.

That next day, I looked up military recruiters in the yellow pages telephone book (the book of information before Siri, the Internet, or ChatGPT) to see what I needed to do to join the military.

"Let Jesus take the wheel."

The Call to Adventure
Will the hero meet the challenge?

CHAPTER 2

MY STEPS ARE ORDERED

"Not by might, nor by power, but by my spirit."
Zechariah 4:6 (KJV)

The 1980's was a period of reorganization for the U.S. Army. The transition to an all-volunteer force placed greater emphasis on training and technology. By 1989, as the Cold War neared its conclusion, Army leadership began planning a reduction in strength. Any time there is a reduction in military strength—strength meaning personnel—a requirement of human resource and supply specialists are needed to help service members transition, and a massive inventory of military equipment is mandated. 1983 was a year of global events, marked by Cold War tensions, technological advancements, and significant events such as the U.S. invasion of Grenada, the first flight of the space shuttle Challenger, and the deployment of U.S. cruise missiles in Europe and the Soviet Union's withdrawal from arms talks in Geneva. The 1980's also was when the Army decided to allow both men and women to train together in the same units. We had four platoons; the men had quarters in one half of the building, and women in the other half. We would all meet at the same area for formation, but the men were in first and second platoon and women in the third and fourth platoon all making one company.

My desire to serve began as a young child, recognizing the need to be a part of something bigger than myself. I wanted to serve because my mom would express the importance of knowing what is going on around you. My parents never served but they raised four children that joined the armed forces. My military service was encouraged by my siblings, who are Marines. After so many things happened after I graduated high school, I decided to follow in their footsteps and become a Marine. I enrolled in the Marines pre-basic training, but during that time a young female Marine recruit had returned from Marine basic training and shared stories about the treatment of women and her training. After hearing her story, that made me uncomfortable, and I decided to consider other services. Since I had no desire to join the Navy, and the Air Force had a waiting list, I spoke to an Army recruiter because the television advertisement during that time was "Be all that you can be in the Army." I discussed joining the Army with my mother, because so often she had shared the story about her brother, Uncle John, who had served as a Soldier during World War II, and how he helped support her with his basic allowance when she was a teenager. I made the decision to follow in the footsteps of Uncle John and joined the United States Army.

When I initially joined the Army, I enlisted through the delayed entry program in December of 1982 at the age of 22. Basic training was called initial entry or basic combat training and lasted eight weeks. It was broken down into five distinct phases.

- **Phase one**: *reception.* Introduction to military life.
- **Phase two**: *yellow phase.* Focused on discipline, physical fitness, and basic military skills.
- **Phase three**: *red phase.* Intense physical and mental challenges emphasizing combat skills and teamwork.
- **Phase four**: *white phase.* Advanced training in weapons handling, tactics, and field exercises.
- **Phase five**: *blue phase.* Final phase preparing recruits for active duty, culminating in graduation.
- **Graduation**: The transition from recruit to Soldier.

Me when I first entered the U.S. Army.

In February of 1983, I entered the U.S. Army and shipped out to basic training at Ft. Dix, New Jersey. The time spent on the delayed entry program did not yield any time served, it only marked your entrance into the service. When I arrived at the base for basic training, I was met with four feet of snow. The other trainees and I were ordered to shovel the entrance way and sidewalk located outside of the building, where we were processed through the reception station, leaving behind our luggage and personal items.

I could sense that at this stage of my journey, there was a spiritual and physical transformation happening. I knew quickly that prayer and my relationship with God would be the only thing that would get me through this new environment that was unfamiliar to me. We received instructions loudly from our drill sergeants, got processed through personal, medical, supply, and finance, and were assigned barracks and company assignments. My company was bravo, third platoon, and the motto was *third heard*.

THE INVISIBLE HAND

After 10 hours of processing was completed, we were ordered to gather the two duffle bags of military supply and ordered outside where several cattle cars were waiting to gather us like livestock. We were told that we had 30 seconds to get on the large metal vehicles designed to carry animals, with no seats, or seat belts. The only thing we had to hold on to were long silver poles in the cattle car. The drill sergeants were lined up once we arrived at our company area and the barracks. More drill sergeants were yelling, "Pick up those bags. Your mama is no longer here to pick up after you." "You aren't in Kansas no more." Hustling us forward, we had less than 15 seconds to get off. I quickly adapted by placing one duffle on my back, and the other in front as I got off the cattle car. My recruiter at the recruiting station prepared me for this. I knew this was mental, and emotional psychology so, I was prepared for this, but it still didn't stop me from feeling fearful.

Once in formation, names were called, and we were assigned rooms. Sitting in our designated areas, we were told to line our duffel bags against the walls and sit on the floor. After 15 minutes, Drill Sergeant, Staff Sergeant (SSG) Rivera entered and made it clear: "You are no longer women, girls, or ladies. You are Soldiers. From this point forward, you will be treated as Soldiers." At that moment a tear escaped my eye, but I quickly gathered myself. Being a woman in uniform came with its challenges. Female Soldiers were being tested and tasked differently than the males. A painful reality that many endure in silence. I realized I needed to stand strong and refuse to let bigotry define my service. Success can only come from doing the right thing, showing up, following orders, thinking critically, being responsible, maintaining appearance, hygiene, manners, and behavior. I embraced basic training, recognizing it as an opportunity to be part of something greater than myself. That mindset carried me through my entire career.

The mindset of a soldier is deeply ingrained in a calling that often cannot be explained. There is an unshakable desire to serve, to step forward when others hesitate, to sacrifice everything for a greater cause. Where does this come from? I may explore that in another book, but for now, I understand that it is not a simple decision; it is a lifelong commitment.

To those considering the armed forces as a career, you should first research each service. This will help you better understand the mission of that service, so you know if you are a good fit. Not all military services are created equally. Military service is often held up as a model of discipline, tradition, and honor, and rightfully so. It offers a structured environment where purpose, duty, and commitment are emphasized at every level. There is a rhythm to military life, but even within this revered institution, perfection does not exist. The structure that brings order can sometimes feel rigid and unforgiving. The very systems designed to promote fairness, and readiness can occasionally overlook individuals, especially those who challenge the norm or come from underrepresented backgrounds.

Not all service members experience the military in the same way. Some face discrimination, bias, or unequal treatment despite the values the institution upholds. To serve is still one of the noblest callings. But acknowledging the imperfections of military life does not diminish its value, it strengthens it. It creates space for truth, for reform, and for healing. Military service may not be perfect, but it is meaningful. And within its imperfections lies the opportunity for growth, not just for the military, but for everyone who answers the call to serve. When thinking about joining the military, know yourself. You must understand it is more than just a career path; it is a commitment that demands discipline, sacrifice, and courage. It means stepping into a life where your actions affect not only yourself but the lives of others, where you may be called to serve in difficult places under difficult conditions. It is a decision that requires honesty with yourself about what you are willing to give and what you are willing to endure. Then you must ask yourself the question: Am I ready to make this choice to dedicate myself to a mission greater than myself?

THE INVISIBLE HAND

I can tell you from my own experience this will not be an easy journey no matter what your age, race, gender or faith. You will be challenged mentally, physically, and intellectually. There are things you will see and do that will stay with you forever. Can you make it and be successful? Yes, but know that serving will change your life in ways you may not foresee and you will be a part of a camaraderie that will be with you always. You will be proud of your accomplishments and the lives you change along the way. I chose to serve. My love language has always been caring, supporting, mentoring and sharing. I know to inspire you must connect on a deeper level, something that comes with being a natural leader.

We often live life forward, but we only understand it when we look back. What you despise right now may end up being one of the best things you could have gone through. I have gone through many difficult things in my life just as I am sure many other people have, but the experiences I had serving this country truly have made me a better person. I was ready for the next steps in my journey. During those eight weeks of basic training, I felt like I was the leader and big sister of the group because I was 22, older than most of the Soldiers. It took every lesson I learned during my school years as a student-athlete to overcome the challenges I faced in basic training, the value of discipline, teamwork, and pushing beyond limits when the game was on the line. Early mornings, long practices, and the drive to stay focused taught me that success comes from resilience and sacrifice. Those same lessons translate into military service where the mission depends on your ability to trust others, stay committed under pressure, and give your best even when you're tired. Being ready to serve in the military meant carrying forward what I learned on the field or court: that strength isn't just physical, it's the mindset to keep going for something greater than yourself.

When I found myself in tough situations, I was able to reach into my "toolbox" and pull-out strategies to cope and keep going. We had to remember many things, like the Warrior Ethos, Army

Values, Soldiers Creed, special musical phases related to serving in the Army called Cadence, and the story of Audie Murphy, a highly decorated Soldier during WWII. We were given classes about the mission of the Army. The three General Orders were emphasized every day.

- **General order number one:** "I will guard everything within the limits of my post and quit my post only when properly relieved."
- **Number two:** "I will obey my special orders and perform all of my duties in a military manner."
- **Number three:** "I will report violations of my special orders, and anything not covered in my instructions to the commander of relief".

The most important class in Basic training was the Basic Rifle Maintenance (BRM) class. This was the most important thing to me because it could save your life and help you save others. BRM taught everything you need to know from the components to the functions of an M16-A1 rifle, M60 and Grenade Launcher. In that era of time, we also had to qualify on the Grenade ranges with live grenades, set, aim and arm a clamor mine. Before graduating basic training, you would have to break each weapon down, put it back together, perform safety checks, then qualify. Skills that would surely come in handy if you ever must serve on the battlefield.

Those were some long hard weeks, early mornings and late nights. I remember being called out to the fields before formation, late in the evening and weekends, for grass drills. The Drill Sergeants would smoke us (break down every muscle in the body) till we couldn't stand, Soldiers would be vomiting and fainting during the drills. Then, the next morning, we went back out doing more physical fitness. But ultimately, I got the hang of it and was able to keep up with all the changes. I was glad I was able to stay focused on the things I needed to do, which was not the case for some of my battle buddies. People had problems, and family drama that kept them up at night, and out of focus during the day. Some women were even coerced into providing personal favors and taken advantage of in the environment that was supposed to be safe. Some of my friends in my platoon would talk about being exploited by Drill Sergeants and other male Soldiers in our company area. It would always start with little gestures aimed at seeking sexual favor, telling explicit jokes, then advance to touching and groping, especially if you were failing at something, like physical fitness or weapons qualifications. One of the Soldiers left everything behind, physically ran across the post to the cab station area near the base exchange, and left. I never found out the true story and you better not be caught asking questions. The rumors were flowing that she was sexually assaulted.

It was not easy for some, but thank God for favor. I did not have that experience personally, and I felt bad for those that had to suffer that emotional and physical trauma. I made it through all the training, the 10-mile road marches with 40 lbs on my back, the Gas chamber, low and high crawling, countless obstacle courses, and the disgusting portable toilets and sometimes the absence of restrooms we had to endure while in training. By the time I completed basic training, I had worn the platoon stripes for squad leader twice, and received a letter of recommendation from the company commander and my Drill Sergeant for my teamwork and leadership. I was promoted to my first rank, Private: one stripe, mosquito wings they were referred to. I was proud of my accomplishments, marching down the avenue carrying that last phase final stage blue flag, singing all the military cadence, knowing in my soul that it was grace that brought me through.

Transition from Basic Combat Training (BCT) to Advance Individual Training (AIT) was why we all worked and trained so hard. Before departing basic combat training, you had to clean your area, turn in linen, gather belongings, sign out of your assigned unit, and travel to the next schoolhouse, if it was in a different location from your basic combat training. There you would learn what the Army enlisted you to do: your military occupation specialty (MOS). Advanced Individual Training phases in 1983 were structured into: initial training in a Soldier's specialty, more privileges such as off-post passes and civilian clothing, and additional freedoms and responsibilities resembling active-duty service. I was assigned to Ft. Lee, Virgina, now known as Ft. Gregg Adams. Ft. Lee is a key Advance Individual Training location for: Ordnance Mechanical Maintenance School, Quartermaster School, and Transportation School.

My first military occupation specialty was 76V (Supply Specialist), which was only four weeks long. The 76V, military occupation specialty involved maintaining stock records and inventory, operating material, handling equipment, handling storage, packaging, shipping, managing property disposal, and redistribution evolution of Army military occupation Specialty Structure.

Ft. Lee had the best dining facility. The cooks there had been preparing meals for Soldiers for over 30 years. The menu served some sort of soul food every day. We had to do physical fitness twice a day just to maintain our weight. The dining facility was the place to connect to other Soldiers and learn about other military schools. By 1993, Ft. Lee continued to be a major Advanced Individual Training location for the Ordnance, Quartermaster, and Transportation schools. Over time, military occupation specialty classifications were reorganized, and the supply specialties military occupation specialties were merged into one military occupation specialty skill identifier 92A, which remains the same today.

After Advance Individual Training, we were allowed two weeks of leave to go home before reporting to our next duty station. I went home, found my husband, and served him with divorce papers. He signed without hesitation, then I returned to Texas to start fresh. The journey I had started had only just begun, but I knew that I was on the right path.

THE INVISIBLE HAND

There's a funny story that goes with my new assignment. When I arrived at Ft. Hood, I met up with a friend I had made during Advance Individual Training at Ft. Lee. We went to visit her unit, which was the 1st Cavalry Division. I did not know anything about reading my orders and understanding what unit I was assigned to. We were with the same military occupation specialty; she invited me to come with her to her unit and in my naivety, I went. She took me to her barracks, and assigned me a bed, locker, and unit patches. I got up every morning following the unit's routine and went to work with her.

Me with my parents.

After a couple of weeks, a Sergeant (SGT) approached me and asked, "Where do you belong?" I confidently told him, "Right here." He looked at me and said, "No, you don't. You belong across the post. Get your orders, get your stuff, and get out of my barracks." It turned out that I had signed into the wrong unit. I had no idea about all the logistics involved, and no one had explained to me what I needed to do when I arrived at the Army base. I didn't know about the in-processing procedures or the welcome center where I could get the information I needed. It was a tough lesson about blindly following others and not asking enough questions.

Fortunately, the Sergeant sent me in the right direction, and I eventually made my way to the 13th Support Command where I was originally assigned. I supported the Aviation Brigade that managed supply parts for the Army helicopters like Blackhawks, and Chinooks. It was a humbling experience, but a good lesson that taught me the importance of being informed and prepared.

"God can put his finger in there and mix it all up, and make change happen right now."

The Refusal of the Call
The hero resists the adventure.

CHAPTER 3

CHALLENGES, CHANGES, AND CHOICES (THREE CS IN LIFE)

"To everything there is a season, and a time to every purpose under heaven."
Ecclesiastes 3:1 (NKJ)

F t. Hood is one of the larger Army Installation in the United States. It is known for the training location for WWII tank destroyers. Home to over 70,000 Soldiers and their families. I was assigned to a two-man room in a nice high rise building on post. I met several new friends and worked in a warehouse that had many shelving units filled with all the parts maintenance personnel needed to repair and maintain helicopters. It was different there; Soldiers were still wearing the old Army green uniforms. I had been issued the new battle dressed uniform (BDU). Just a small group of people worked in the supply warehouse, but we all had our assigned responsibilities.

I did not have a car, so I made friends that did, so we could travel around Killeen, Texas. Killeen is a small city outside Ft. Hood. We would go to the mall and the movies during our off time and get some good food at the local restaurants. The best times were when we all got together to play spades and listen to music. One of my coworkers was from St. Louis, Missouri which is just across the Mississippi river from my hometown East St. Louis, Illinois. He would talk all the time about how he went home to see his family as often as possible. One four-day holiday weekend, he convinced me to join him and two friends to ride back to St. Louis. I didn't see the harm, I could surprise my mom and dad, but I honestly didn't know how far St. Louis was from Killeen, Texas. I assumed since he went home all the time, he had that drive memorized and timed to make it to Missouri and back to Texas.

It was amazing going through all the small towns and discovering things in areas I had never traveled before. We arrived early that Saturday morning. My family was waiting for me, and it felt good to be home. My mom and dad spoiled me like I was a little girl, cooking all my favorite dishes and the best part was sharing all my accomplishments and achievements from basic training and military school. I proudly shared my letter of recommendation with my family. I remember feeling like I had done something special. My mom told me she was always praying for my safety. I attended church services with my mom that Sunday and the whole congregation gave me a standing ovation for my service; there is no place like home. My mom and dad were always proud of my service and telling everyone they could about their daughter in the Army. I had the best time with my siblings sharing my Army stories since they thought Marines were the only military service. That time was one of my joyful memories.

The four-day weekend went quickly, and my ride arrived Monday to pick me up to return to base. He introduced himself to my mom and told her he would get me back to Texas safely. My mom had prepared fried chicken and potato salad for the road, and said a traveling grace over us. Then we got back on the road, but what a story to tell about our return. The trip really was a two-day drive, but he made it in less than 24 hours by only stopping for gas. We left St. Louis later than we should have on the return, and he was speeding and driving recklessly in and out of lanes to get back for formation at 6:00 AM, Tuesday morning. Once the day turned into night, I could tell he was getting tired, so he turned on the air conditioner full blast to help keep him awake, and he had all the

Challenges, Changes, and Choices (Three Cs in life)

windows down. That wind and air were cold, and I was freezing, tired, and afraid to fall asleep, so I started rubbing my hands and arms to generate some warmth while silently praying. I asked him if he wanted me to drive so he could rest for a few hours. He said, "I take a lot of back roads to make up time and you won't know the way," so I sat quietly.

We were on the trip for about 10 hours when we arrived in one of the small towns he assumed he could speed through to get back on the highway. At the last intersection before the highway entrance, a police officer clocked us for speeding and lit us up with the siren and his red, white, and blue lights. He proceeded to chase us. My friend said, "I can't get stopped; I'm pushing it to the highway." I didn't ask him why not, because I assumed he had some speeding violations. I was in shock. "We can't outrun the police," I said. All I could think about was all those movies I had seen about being black and arrested in small towns. I was holding my breath and praying. He said, "I am going to outrun the police and jump off one of these exits and stop to hide."

Back then, truckers used citizens band (CB) radios to communicate. My friend had one of those two-way radios in his car. He got on the microphone to ask the truckers for help. The truckers answered and started giving directions. He began maneuvering that vehicle in and out of the traffic and between the 18 wheeled trucks to avoid getting caught, all while talking on the CB radio. The truckers would block us in so the car couldn't be seen which allowed us to get off the highway before the police could catch up to us.

We drove for about five minutes until the truckers gave him the OK to get to the next exit. He got off the exit and shut off the car and lights and we waited awhile before we got back on the highway. That woke us up for sure. I told him he was crazy and that was dangerous putting us in that position. He said, "Don't worry, I do this all the time." I looked at him and shook my head.

THE INVISIBLE HAND

We eventually made it back safely to the post about 15 minutes before formation. I jumped out of that car, kissed the ground, and thanked God for a safe arrival. I ran to my room to get changed into uniform to meet formation. The First Sergeant (1SG) was serious about being prompt for formation, and he would not hesitate to give extra duty if you were late. He would say, "If you were not in the formation 10 minutes prior to the first call, you are late." I made it to formation with just two minutes to spare. I never rode with that same guy again and invested in my first car soon after.

Training on our wartime mission at Ft. Hood was consistent. Prior to moving out to the training site, we would have to lay out field gear in a specific order for easy inventory. Once the First Sergeant inspected your gear, you could pack it up and return it to a stored location in your room which was on top of the wall lockers near your assigned bunk. We were in the field every 90 days performing different training exercises. At all times of the night, you would hear those tanks rolling along the trails in the field. They used to call them the tracks of death. Before every training event you had to sit through a safety briefing. Your leader's safety brief always included the terrain and parameters that you had to stay within to stay safe. Soldiers will always say people can get killed if they get caught in the path of a tank. They were called rollover accidents. A Soldier lost his life in July of 1984 due to a tank rollover.

I was not one to go off path, because it was always so dark outside. We would have to walk a long distance to use the bathroom, or latrine, they call it. Once you got to it, you hoped it had been cleaned after the last exercise, because they were gross and disgusting. You could smell it before you saw it. Due to my school sport years, my mother had always taught me to carry a small pack of tissues and wet wipes. I still have those same items in my purse and car just in case. Those wet wipes and tissues have saved me many times. Showers in the field were slim to none. It often depended on your scheduled duty if you got a chance to shower. I would use the wet wipes and the water from my canteen to clean up.

Challenges, Changes, and Choices (Three Cs in life)

You were supposed to always move in pairs. My roommate was my battle buddy, and we would always move out together. When we were not training, we were placed on a duty roster to serve some kind of duty. Guard duty, Kitchen Patrol (KP) duty, or Charge of Quarters (CQ) duty. It was always a two-man duty, one non-commission officer, Sergeant and above, and a junior enlisted Soldier (Private-Specialist).

Every Soldier has their everyday duty uniforms and spit shined boots, but you had one uniform and a pair of super shined boots for Soldier boards and reporting to duty. The duties were 24-hour shifts, and you would have to report to formation in the special duty uniform and highly-shined boots at the company headquarters at 4:30 PM after your workday. I am no stranger to these duties because we pulled these duties during basic combat training, and Advanced Individual Training. The duties we never hard, just long. Usually you would be tasked to answer phones, sign service members in and out for leave, guard the front entrance to the barracks or the company headquarters, be a driver for the headquarters overnight in case of emergencies. There were always additional duties, such as sweeping the floors, and cleaning the restrooms. You better not be caught sleeping on duty because of your first general order, which you learned in basic training. If caught, you would be punished by the First Sergeant and Company commander. You could lose pay, or a rank, and even worse, you would be put on the duty roster every night for 30 days straight.

After your duty, you would receive the next day off. The First Sergeant would come out to formation, and do an in ranks inspection, and duty assignments for the night. He was always looking for ways to help you learn and improve your knowledge of the Army history, traditions, and current events. This knowledge would help you during your Soldier boards to earn rank and rewards. The First Sergeant had its own award program where you could earn two days off, the duty shift and the next day off, as a reward. He had prepared a list of questions from the Soldiers study guide that was given to you in basic training. I was always studying that guide because I wanted to be all I could be, so when Soldiers got smoke breaks during training or class I would pop out my book and study. Smoking was one thing I never had the desire to do.

During formation he would make us compete against one another. He would get in your face to ask everyone questions, and you would have to answer correctly to win the two days off. If you missed, you had to do the overnight duty. I loved the competition because iron sharpens iron. I lost a few times, but then I got better and more determined to win. Not everyone can handle the pressure of losing. One time I won against this white male Staff Sergeant, he was not happy. After the First Sergeant released us he ripped his ID dog tags and stormed off. The next morning he was found on the guard duty post deceased; he used his weapon and committed suicide. I never forgot that. To this day I wonder if it was because he couldn't handle losing, or losing against a Jr. Soldier (E1-E4), or a black female. Things happen while you serve that you take with you always. I just remember that trouble doesn't last always.

While I was at Ft. Hood, I worked with a peer that I gravitated towards. After a while, we started dating, and eventually got married. Three months after we got married, we both came down on levy (overseas orders). We later found out it was a tour of duty in Germany. I did not have a good experience with Germany because my first husband situation, but I accepted the assignment and began going through the process to get ready for my permanent change of station. I was determined to be successful and make the best of the assignment.

During that process, I discovered I was pregnant. You could not change duty stations while pregnant, because once the pregnancy is confirmed, you are issued a pregnancy profile until six weeks after the birth of the child. I discussed this with my spouse, and this led to one of the hardest decisions I had ever made. I was informed that I had two choices—to give up my child for a year to a family

Challenges, Changes, and Choices (Three Cs in life)

member to stay in the military, or accept Chapter 8 (voluntary separation), receive an honorable discharge, and leave the Army no later than seven months into my pregnancy. Chapter 8 provided medical care for myself and the birth of my child up to six weeks after the birth, but I would not receive any more pay.

The idea of being separated from my son was unbearable. Without question, I made the tough decision to take the discharge, and leave the Army, opting to join the Army Reserves after giving birth to my child. I was discharged seven months into my pregnancy and returned home to East St. Louis. My husband departed for his duty assignment in Germany with the understanding that after I gave birth, he would send for me and our child. He had to get permission from the command called a (command sponsorship), for me and our child.

My son was born December 2, 1984. My husband never sent for us. I tried to contact him, but he was not returning my calls. It was nine months after my son was born when I saw my husband again. He called my mother and asked if my son and I were there and could he come visit. That was a tense time in my life. I called my brothers and asked them to come to the house; I wanted to hurt him. All I could think of was, *I gave up my Army career to be his wife, and the mother of his child, and all I received in return was silence.*

When he showed up, my mother said, "Pamela, you are still his wife, let him explain." I was hurt and angry, but I did as my mother asked. I sat down to hear him out. He explained he had gotten in trouble in Germany with narcotics and alcohol consumption, and he was being discharged from the Army because he tested positive for drug usage during a urinalysis. I stood up and hit him in the face with my fist. My brother grabbed me and dared my husband to hit me back. I was thinking to myself, *what am I going to do now? Back in the same position, but now as a mother, no money or job.*

I tried to make it work but it was very hard trying to find work that would accommodate an affordable wage so that I could support myself and my son. My spouse provided $270 a month, which then was the basic housing allowance and only covered the car note. With my mom and dad's assistance, I was able to get on my feet and move into my own place. I had no choice again but to choose divorce. I was not going to raise my son in an environment with chaos, drugs, and alcohol. So many women suffer through marriage for their children or security. I was not going to depend on a maybe. I knew God did not bring me that far to leave me. It was a brave decision to choose peace and a better future. Divorce has taught me who I am, what I value most, and what I won't settle for again. The end of that chapter was not the end of my story; it was the beginning of a new one.

After my divorce, I moved on, holding on to the idea of love. Not just any love, the kind that lasts, the kind I saw in my parents. I didn't give up on it just because things hadn't worked out before. I still believed in building something real, something that could stand the test of time. I was ready for a fresh start, open to love again, hoping that maybe the next relationship would be different.

I followed up with my contacts, Mr. Andrew Finley and Mr. Marvin Ware (both were civilian administrators and Sergeant First Class in the Reserve unit) and reported to the Army Reserve Center in Granite City Illinois. I served in the 624th Engineer Map Distribution Platoon. The mission was to purchase, package, and store maps for wartime missions, and to backfill active Army Soldiers in Germany and Korea if war was declared. At that time, reserve duty was one weekend a month and two weeks a year. Our yearly training was served in Korea and Germany. We traveled together and worked in warehouses overseas with Soldiers on our wartime mission. I volunteered for additional active-duty days to assist the unit with special tasks and events. Since I volunteered for duty so often, it opened the door for me to attend school and earn a second skill identifier. Because I was already performing

Challenges, Changes, and Choices (Three Cs in life)

extensive administrative work, I attended school for MOS 71L, which later converted to 42A. Those skills turned out to be very critical as I advanced in rank. I loved being a Soldier. I knew it was my true calling. I finally began to move forward, getting things together for myself, and my son. I found peace in the pieces.

Sometimes the hardest decisions are the ones that set you free. In the years that followed, I made choices that didn't always work out. I kept finding myself in similar situations, chasing that same kind of connection, but landing in the same pain. So yeah, that chapter was painful, but necessary, and somewhere in that I started healing. God met me in the space between letting go and holding on. He gently showed me that my worth was never tied to someone else's ability to love me right. God began to heal the places I didn't even know were broken, and little by little I stopped chasing what wasn't meant for me. He wasn't just helping me recover; he was preparing me for something greater.

On August 2, 1990, Iraqi President Saddam Hussein sent over 18,000 armed forces to invade and overthrow the Emirate of Kuwait. Under United Nations' support, the United States and other member nations responded by deploying military forces to Saudi Arabia, to aid in the long-term goal of forcing Iraq to withdraw from Kuwait. The operation was named "Desert Shield." In January 1991, military forces began to remove Iraqi forces from Kuwait. The operation was called "Desert Storm." U.S. Military involvement was because President George H. W. Bush, under the advisement of Secretary of Defense Dick Cheney, believed if Saddam Hussein was successful, he would control 20 percent of the world's oil reserves. This was considered the turning point of the President Bush administration.

My unit, 624th Engineer platoon, a 12-man team, was called to deploy August 1990 to support the congressional authorization for the use of force. I remember the news headlines stating the U.S. had declared War. I was among 12 individuals selected for the deployment, and we were sent to the Middle East to assist in the effort. Our team mobilized out of Scott Air Force base in late September of 1990. We arrived in country and were driven to Bahrain. We were attached to the U.S. Navy in Bahrain. Our lodging quarters were off base at one of the Sheraton hotels in town. This was only temporary quarters until the villas being built for us were completed. We had to be transported in plain government vehicles and escorted by a local transport company to work daily in plain civilian clothes to and from work. We would change into our daily brown desert chocolate chip uniforms and wear the special gear for the mission once we arrived at our work location. Our mission was to map the gulf. We were assigned to a topographical detachment to assist with getting the correct maps to our service members because that terrain had not yet been developed as a battlefield combat zone. I had a secret clearance, so I was assigned to work with classified and secret maps. That assignment allowed me to work directly with all military services on special missions.

Me on deployment.

One of my missions was to travel with a commander to his Navy ship in the Gulf with secret maps. The maps were locked in a

Challenges, Changes, and Choices (Three Cs in life)

briefcase, and the briefcase was handcuffed to me. We flew by helicopter to his ship at night. What an experience! When we landed, we met his staff and were escorted on board into his quarters. I was a Specialist, (Jr. enlisted Soldier). I had to be handcuffed to the Commander because he explained that the ship had been out to sea for four months, and tensions were high within the vessel. He told me I needed to be secured to him, because I could go missing on the ship. There are many manholes and spaces on a ship, and if I were lost, it may be days before I was found. I felt this tight grip in my chest. It wasn't just fear, it was the awareness of being vulnerable that I could be violated, disrespected, and dismissed. So I did as he said, delivered the maps, and a few hours later I was back on the helicopter to return to my duty location. That personal experience is something I hold close to till this day.

I was assigned to work with the Navy "Seabees" (Navy Occupation Identifier) for logistical support. The "Seabees" were there to supply construction support to the Armed Forces. They were responsible for all transportation vehicles used by military personnel. When things began to escalate in the area, Command staff wanted to increase safety for our team. The Navy Master Sergeant in charge ordered me to request a vehicle to drive our team to and from work. I went to the area on post to inquire about getting a vehicle. I met a Navy "Seabee" Petty Officer First class, that oversaw the transportation on the installation. He helped me complete the process to get a local foreign driver's license, then he assigned and dispatched my team an eight-seat passenger van and a brand-new four passenger Jeep in white. The papers were still in the windows when I picked the vehicles up. Having the two vehicles allowed me to get mail for the troops and complete local transport for mission essential items.

He and I became friends, remained close, and began to hang out. At first, I wasn't sure if he was staying close to keep an eye on those vehicles, or if he just liked my company. He had been stationed there for a few years and knew the local area well. My work area was a warehouse in the local district of an industrial park. We were located on the back side of the Navy base Administrative Support Unit (ASU), renamed in 1999 Navy Support Activity (NSA).

My team and I would walk to the base every day. We were issued ration and meal cards to purchase food and personnel items from the commissary. We dined during the day at the base dining hall, and in the evening at the hotels. There was never a bad meal. My favorite was Mongolian night. They would have the large cookers, and food was prepared fresh.

Some evenings we were allowed to go into the enlisted club on base for moral welfare and recreation. The famous song then was *Ice Ice Baby*, by Vanilla Ice. Every time I went there, that song was playing. I made friends, and we all hung out supporting each through this emotional event. I remember one afternoon while on shift at work, a unit arrived from the desert to pick up their supplies. The Soldiers, mostly females, got out of their trucks and they were so dusty from the desert sand. When I saw them, I immediately felt empathy. I asked them if I could offer them anything, and they asked for Lysol and cleaning supplies to clean their living quarters and bathrooms. I felt remorse; God had blessed me with so much. I asked them how much time they had and they replied, "Two hours." I decided to share my blessing with them. I took them into the Navy base fitness center where they were able to shower, wash their hair, and a get a hot meal. I used my monthly ration card to get them the cleaning supplies, snacks, and I gave them the supplies we had in the warehouse. They were so thankful, and I thanked God for allowing me to be in place at the time to help them.

Things were going well until January 12, 1991 when congress approved the authorization for military force against Iraq. We were ordered to go get all personnel records updated and we were issued an 8-pack of pills that looked like birth control pill packs, that we were supposed to take every eight hours. The pills were pyridostigmine bromide (PB), an anti-nerve agent pill. The pill was supposed to protect us in the event of a nerve agent attack. Service members were afraid of this medicine because it was not approved by the Federal Drug Administration and was known to have genetic factors, and long-term health effects. There was a lot of conversation among service members that the drug was not safe to ingest. I believe we were intentionally kept in the dark about the side effects, especially the part about unknown birth defects to future pregnancies. One of our reserve members that did not deploy with us sent the medical alert information via email on the pills, and we were warned not to take them.

The platoon leader, a Navy Master Sergeant who we called Broom Hilda, would line us up daily in formation. She would get in our face and threaten us with legal consequences if we didn't take the pill. I was afraid the pill would cause harm to my body and any children I may have in the future. I know for a fact every member of my Army team that deployed with me suffered some sort of medical implication after we returned home. The three female Soldiers never conceived children, the men had medical complications, and the men that had children had medical issues. I have health conditions such as fatigue and chronic tiredness that persist over time, muscle and joint pain without a clear cause, respiratory issues, including chronic cough or shortness of breath. I have experienced gastrointestinal problems, unexplained abdominal pain, and sleep disturbances, including insomnia from all the medicine and shots that were administered before and during my service during Desert Shield/Desert Storm. The study about the effects of the pills is still ongoing research; it has not been concluded that causation of chronic multi-symptom illness that many Desert Shield/Desert Storm veterans suffer from today including myself, was caused by PB.

By January 17, 1991 Operation Desert Storm was in motion and aerial bombing had begun. We were ordered to pack all our personal belongings and load the vans. We had to move into the warehouse to live where we worked because of chatter that the area airport in Bahrain was targeted by Iraqi scud missiles. We all slept in the same area on military green cots on top of the metal containers in the warehouse. We used the warehouse restrooms to take care of our hygiene and wash our clothes in a tent near the warehouse. We had to eat the military "C" Rations (field packaged food) that consisted of prepared, wet foods that were served when fresh meals were unavailable. We were now in full war zone mode. We had 24-hour guard duty and were assigned shift work. We were issued our weapons; no civilian clothes were authorized, and we had to dress in full battle rattle (chemical protection gear and weapon) every day. One of my friends and teammate, Alexis Brice, refused to wear that stuff at night. She would dress in her little lingerie nighties, house robe, and fluffy house shoes until Master Sergeant, Broom Hilda, caught her and said, "How are you going to be ready to fight in your little nighties?" She ordered her to get dressed in her uniform and Broom Hilda took all her clothing and locked them in a foot locker. We all laughed, that was too funny. We lived in that warehouse and those conditions for two months.

I was on guard duty one night, and I was walking and praying. My mother had told me to remember Psalm 46, "God is our refuge and strength, a very present help in the time of trouble." I would walk and pray the entire time and repeat that Psalm. A few days later, very early in the morning, a scud missile hit about 20 miles outside of the Bahrain area. I was sleeping because I had guard duty overnight. When the sirens went off, I jumped up half dressed, dawned my chemical suit and mask, grabbed my weapon, and ran into the warehouse. Everyone was panicking, crying and praying. We walked around for hours afraid to take off our masks. After a while, we were given the all clear to remove our masks. That was a very traumatizing experience. The suddenness and unpredictability had overcome me. It was like a loss of control or helplessness, a real threat to life or well-being, a lasting emotional impact of fear and confusion. I thought about my son, my mom and dad, if that missile had hit us. So glad God was a present help in the sign of that trouble.

Not long after the war had begun, we were informed there was a cease fire in place, and the war had ended. Once we received official notification and were cleared to continue our mission, life was different, fragile almost. I began craving connection, understanding, and a sense of safety. The Navy Seabee Petty Officer would come and bring me food and check on me at the warehouse. Many nights, we would talk through the high wire fence in front of the warehouse or on the phone and relive all the things we were experiencing. He would listen; he was present, and available emotionally, and physically, and suddenly, that presence felt like relief. Sometimes, we confuse comfort with compatibility, and that kind of involvement can lead to misaligned expectations. We had a close bond and before we fully understood our future circumstances, we entered an intimate relationship. For a few short months we were holding on to the moments in that time, not sure if we would live beyond the day. We trusted and believed in each other, but we were both still reeling from trauma. Things were happening that we couldn't understand, and the news reported about the war and casualties that neither of us were in a place to accept.

One night when I was out on patrol I passed out. I was taken to the hospital and the doctor said I was dehydrated. He would run some blood tests, and I needed to rest. A few hours later he came in to give me the results of the test and he said, "You're fine, but you did test positive for pregnancy." I was already grieving, traumatized, healing from my war experiences, and already juggling life. I froze, everything around me felt distant, like the world just slowed down, but my thoughts were speeding up a hundred miles an hour. With that one result, a storm of questions started pounding in my mind. *What am I going to do? What does this mean for my future, my dreams, my relationships? What am I going to tell my family?* I had obligations and responsibilities at home.

When the news was shared with the father of my child, that moment was heavy. There were promises made, but they never manifested. I had to leave the mission and return home. I left Bahrain two weeks later and reported to Ft. Leonard Wood, Missouri. I did not tell my family right away; I had to deal with my own situation. I could not return to home military base until my unit returned to demobilize and process out of the installation to return home.

I had a few months to adjust to my new truth. Was I embarrassed? Yes, and I made a decision that changed the dynamics of my life. My unit returned in April 1991 and I was processed out of the Army on April 4, 1991. I received my second Chapter 8 (an honorable discharge) and was granted the same benefits as I received with my first son. I went home and told my family. It was hard, but I was not ashamed. It happened and I accepted it and put things in motion to prepare for the birth of my child. I had contact with the father up until my fifth month of pregnancy when he admitted he could not keep the promises he had made. I had no words. Emotionally devastated, it was more than being let down; it created a unique kind of ache, it severed the unspoken expectation. I felt neglected. It definitely shook my sense of security and caused me to question my judgment, but I had to get back on track. I now had a greater responsibility, and it was time to move forward and get realistic. There were things I had to release and versions of myself I had to embrace and give myself the strength to turn the page.

After I gave birth to another beautiful son, I returned to my employment at a computer data firm in St Louis, Missouri. Within a few months, I received a call from my Reserve unit administrator Mr. Andrew Finley. He asked me what I was doing and if I wanted to return to the Reserve unit. I didn't know that my Army contract had expired when I accepted Chapter 8, honorable discharge, and my six-year obligation was complete. He said they had annual training coming up and they were going to Seoul, Korea, and he sure could use my help. He said he remembered how much I loved being a Soldier, and I was honored by his comments and the offer.

Challenges, Changes, and Choices (Three Cs in life)

However, I had no idea what I would have to do to rejoin the military. I learned I had to go through the entire process again, take the Armed Service Vocational Aptitude Battery (ASVAB) test, complete the full medical physical, go through an interview. It took every ounce of determination, and perseverance I had, but I passed everything with the help of my elementary math teacher, Mr. Johnson, who worked with me in the evenings to get my math skills refreshed.

I signed a new contract and rejoined the Army Reserve. From that point on, I made a promise to myself, I would never get off active duty again unless I was ready. It took everything I had to get back, but it was worth it. I wasn't going to let anything stand in my way anymore. I had a new mission, a renewed sense of purpose, and a determination to stay in the service I loved. I had learned so much, and it was time for me to keep moving forward.

Me with my boys.

Healing starts when you stop trying to rewrite the past by authoring your future. My son's father did finally begin child support for his son four years later. Throughout his childhood years, there was fragile rebuilding, role confusion, boundaries, hurtful moments and patience, and now a possibility of healing. He is a part of his life, and he continued financial support through his college years.

I continued my reserve duty and served 11 years. I was always looking for better opportunities to support my family. My older brother Michael worked for the National Park Service, and he informed me that during the summer the Park Service would be hiring for a permanent government position. Thanks to my brother and my veteran status, in May of 1993, I was able to secure a position with the National Park Service at the St. Louis Arch as a Park Ranger. Again, I reached back into that "toolbox" I had cultivated over the years and put that knowledge to work.

All those experiences helped me get to a place where I could support myself and my family while continuing my service in the reserves. I was able to go to school for both College and Army Leadership courses. I advanced in rank and my leadership skills. Life was still hard, I had to work four different jobs, seven days a week, to make ends meet. Some nights, I cried in the bathroom. Not out of weakness, but because the weight of it all needed an exit. But I never let the tears last long. There was always something that had to get done. Someone who needed me more than I needed rest. Working multiple jobs didn't make me less than. It made me resilient. It taught me how to stretch time, how to breathe through chaos, how to keep hope alive on empty.

I would stay with my mom and dad for some days, when the shift work was too close for me to get the children to school and make it to work on time. One morning, my mom woke me up to get ready for work and she asked me what day it was, and I could not tell her. She said in her soft voice, "Pamela you can't continue this path.

Challenges, Changes, and Choices (Three Cs in life)

You are not getting enough rest. I am worried about you. Find one job to support you and your children."

In February of 1996, I was afforded the opportunity to serve as an augmentee for a 179-day tour on active duty in Ft. Dix, New Jersey. I thought this was my opportunity to work one job and be able to provide better for my family. The assignment was to support the Bosnia Joint Endeavor mission on the base. I completed the requirements, and in May 1996 I was en route to the place where I started my Army career. My mom and dad kept my sons during the summer so I could participate in this mission. I was now a Sergeant, a Noncommissioned officer. I worked at the headquarters building with Troop Issue Subsistence Activity (TISA). My duty assignment was to order supplies for the dining facility and the minimal security prison located on the military installation. I had an awesome time, met some great friends from both New Jersey and Philadelphia, and we had fun exploring the East coast.

I was met with a challenge that I didn't expect. While serving at Ft. Dix, I was walking down the avenue past the Noncommissioned Officer Academy, returning to work from lunch one day, and I was called out by a Sergeant Major (SGM). The Sergeant Major yelled "Soldier, where are you headed?" I replied, "Work, Sergeant Major." He signaled for me to report to him. I hurried to his location and stood at parade rest (a formal position assumed by a Soldier in ranks, in which he or she remains silent and motionless). He said, "I see that combat patch on your shoulder, what unit did you serve with?" I replied, "The 624th Engineer Platoon." He said, "Wrong answer. I am giving you an attaboy (encouragement). I want you to research that combat patch and meet me back here at 1700 (5:00PM), with the correct answer." I replied to him, "Roger that, Sergeant Major."

I learned that I was assigned to 20th Engineer Brigade of the XVIII Airborne Corps in North Carolina. This brigade was the higher headquarters of the 624th Engineer platoon I deployed with during

Desert Shield/Desert Storm. I returned at 1700 (5:00PM) and reported the information, he said, "Right answer, now you know."

He asked a lot of questions about deployment, career, family, and how I ended up at Ft. Dix. I explained that I volunteered as an augmentee to support the Bosnia Joint Endeavor mission. He asked if I wanted to come on duty full-time. I responded, "Yes, Sergeant Major." He asked if I knew about the Active Guard Reserve program. I responded, "No, Sergeant Major." Sergeant Major replied, "The Army needs more dedicated Soldiers with your experience and I am going to help you get back on duty." He told me to come back to complete the packet, offering to help me secure an interview for a position at the Noncommissioned Officers Academy on Ft. Dix. It felt like another one of those moments where I had to reach back into that old toolbox, the one that taught me to be ready to receive whatever God has planned for me.

Me with my boys, about 1997.

Challenges, Changes, and Choices (Three Cs in life)

I went to work, figuring out how I could complete the packet and get back to Active Duty. It wasn't an easy task; I'd been through so much, managing several jobs at the same time, broken promises, shattered friendships, multiple marriages, and torn relationships. Keeping a roof over my children's heads, and paying bills alone was never easy, and I needed a new beginning. I was ready to embrace the role of Soldier with more resolve than ever, especially since I had two African American boys. The statistics where I come from weren't encouraging statistics that showed the likelihood of them going to jail or getting killed was far greater than the chances of them going to college. Looking back, it wasn't an easy journey nor was it easy to leave my family's village of support, but I returned home to my reserve unit in October and applied for the program.

In February 1997 I received a letter offering me a position in the Active Guard Reserve program as a Medical Logistics Specialist at Ft. Meade, Maryland. I was able to get back in as "needs of the Army," which meant I had to take whatever military occupation specialty that required immediate backfills. That's how I earned the skill identifier to be in medical logistics, a choice I never expected, but I accepted it and made the best of it.

"Make a difference wherever you go."

Meeting the Mentor

A teacher arrives.

CHAPTER 4

THE SOUND OF MY YES

"Then I heard the voice of the Lord saying, 'Whom shall I send?' And I said, 'Here I am send me.'"
Isaiah (NKJ) 6:8

In March 1997 I received orders and reported to Ft. McCoy, Wisconsin for a five-day course to attend The Active Guard Reserve Entry Program. This course provided an orientation to roles, responsibilities, entitlements, and benefits. All those days of extra duty working in the reserve office had paid off. A lot of the information and processes were familiar. Once I completed the initial entry training, I reported to Ft. Sam Houston, Texas to attend the five-week Medical Logistics course. Back to barracks living, details, daily formations, marching, cadence calling, group physical fitness, and dining facilities (mess halls).

I reported to the unit company office and was in process into the school course. I was considered prior service, so I was assigned to a small two-man room in a building near the main company. The building housed about 50 prior service or Military Occupation Specialty reclassification (new MOS) individuals, and it was coed. It was two stories; men were assigned to same gender rooms on the second floor, and females assigned to same gender rooms on the first floor. There were no latrines (bathrooms) in the assigned rooms; we had one large latrine assigned to each floor and one laundry room per floor. We had assigned times to use it and clean it. In every troop building on the installation, it is the tenant's responsibility to maintain cleanliness. The platoon leader or squad leader would maintain a duty roster with assigned dates and times for teams to clean the bathroom, and build common areas (halls, bathrooms, laundry rooms) both indoors and outside.

I graduated in the top 10% of the course and received orders from my first Active Guard Reserve permanent duty station, Ft. Meade, Maryland. I was assigned to the 410th Medical Center, Theater Material Management (TMMMC), located in Upper Marlboro, Maryland. Our unit operated out of a large Army Reserve Center that housed several other units. The mission was clear: support and execute critical medical logistics operations. Despite being a Sergeant, I stepped into a range of high-responsibility roles. While I was getting settled into housing and learning my way around the base, I met a familiar face, someone I thought I knew from my hometown. She introduced herself and said, "I'm Deborah Johnson." I'll never forget the day we met in the base exchange. She looked over and said, "Are you from East St. Louis?" Just like that, we clicked. It was like finding a piece of home in the middle of everything. From that moment on, it was the start of a lifelong bond. Even after all these years, we are still family, and the bond just gets stronger with time.

I served as the Noncommissioned Officer in Charge of the Material Management Division, Stock Control Supervisor, Property Accountability Officer, Records Manager, and assisted the Unit Administrator. It sounds like a lot, and it was, but in Reserve units, especially at the Detachment or company level, it's common for full-time personnel to wear multiple hats due to limited staffing. The Reserve Soldiers drilled once a month, assisting with training, administration, equipment and vehicle property maintenance. Once a year, we conducted annual training at Ft. Detrick, Maryland under the guidance of the United States Army Medical Materiel Agency (USAMMA).

Their mission, delivering medical materiel readiness and sustaining global healthcare operations, became part of my DNA. I absorbed everything I could, learning through hands-on experience and direct mentorship. I owe a profound debt of gratitude to Sergeant Major (Ret.) Milton Watford and Major (Ret.) Shelia Watford, two leaders who truly saved me from myself. From my very first day on duty, they took me under their wing and provided the sincere mentorship I needed to navigate the complexities of leadership. They didn't just point the way; they shaped my entire direction, teaching me how to lead with integrity while providing the opportunities necessary for my growth. Most importantly, they acted as a shield, protecting me from the professional wrath that often accompanies small unit organizations. I can say with absolute certainty that I would never have reached this level without them.

In that fast-paced environment, I learned to streamline operations and balance an incredible workload. My commander and executive officer often expressed their appreciation for my contributions, emphasizing that my efforts were instrumental in the unit achieving its operational and training goals. These sentiments were reflected in my Noncommissioned Officer Evaluation Report, which became vital steppingstones for career advancement.

After ten months, I became eligible for promotion consideration to Staff Sergeant under the Active Guard Reserve program. When the board results came out, I was considered, but not selected. It was my first look, and naturally I was disappointed. A senior administrator in the building, Sergeant First Class Debbie Morant who had just returned from deployment, offered valuable insight. I didn't have enough time on duty as a Sergeant. Her advice was simple, "Stay focused, do great work, further your education and get enrolled into college, and build a solid military knowledge foundation." I took that to heart and enrolled into University of Maryland under a general studies degree program. I completed two classes, one in English and the other in literature, earned six credits towards my degree, and put the time in at work to get a better understanding of the Army and my responsibilities.

When the board convened the second year, I had more time in my current rank, and more education to add to my board file. My hard work paid off. I was considered, selected, and promoted to Staff Sergeant. By the end of my second year, circumstances changed dramatically. The full-time Unit Administrator Ms. Kaye resigned, leaving me as the sole full-time staff member for the entire unit. The workload doubled overnight. I relied heavily on the support and guidance of full-time administrative specialists from the neighboring units of the building to ensure that our personnel matters, especially pay and training, didn't fail. Then came the biggest challenge yet: the unit received orders to deactivate by the year 2000. This meant every piece of equipment, vehicles, computers, furniture, medical assets had to be accounted for and turned in. I rolled up my sleeves and got to work, leaning into every skill, contact, and ounce of experience I had. It was a massive task, but I got it done successfully, deactivating the unit within the year.

It is standard practice for a service member to receive an award for a job well done during a tour of duty prior to a permanent change of station. One day, while I was outside placing placards in preparation for the upcoming deactivation ceremony, making sure everything was correct for visiting officers, I was approached by a woman in civilian clothes. She began asking questions about the unit, how things were going, and the status of our closure mission. Assuming she was from higher headquarters doing a routine site visit, I stayed ready and professional. I walked her through everything, explaining our progress, challenges, and all that I had personally taken on. To my complete surprise, the woman who had been asking questions turned out to be a Brigadier General (BG). Brigadier General (BG) Carol Kennedy from the higher headquarters returned later the next day in uniform, proudly wearing her general's belt and buckle (a symbol of rank and pride), to attend the deactivation ceremony. During the ceremony, she shared with the audience how she had been doing an informal recon to locate the unit and saw me actively engaged, leading from the front. She said she was deeply impressed by what she witnessed and had never been prouder to approve an award for such an outstanding job done in support of the command. She insisted on personally presenting my award. That day I received my first Meritorious Service Medal, a symbol of everything I had endured and accomplished as a Staff Sergeant with the 410th Medical Center. My mother, aunt Lilian, and cousin Janice from Chicago, Illinois were in attendance. After the ceremony, my mother looked at me with tears in her eyes, and simply said, "I am so proud of you." It was truly a day to remember.

While completing the final steps to deactivate the unit, I received my next duty assignment. I was disappointed when I read the orders: I was assigned to go to Fairbanks, Alaska. I contacted my career manager immediately to confirm whether this was a mistake. She told me, "No, that's correct, there's nothing else available at this time." As the good Lord would have it, a friend in the building heard of my assignment and stepped in to help. She introduced me to the Command Sergeant Major (CSM) Paul Sinclaire of a unit at Ft. Meade, who just happened to have a cousin living in Alaska, Mrs. Jennie Worrells. The Command Sergeant Major told me not to worry; he would call her personally and let her know I was coming. He said, "She'll take good care of you and your family." And that she did. Jennie introduced me to many people in Alaska including my church mom and dad Pastor Cleveland and Annie Bartley who continue to pray for me and my family till this day.

From the moment I arrived, Jennie and the Bartleys looked out for us, helping my family and I get settled in and navigate the very different transition. Their kindness and support made all the difference. To this day we remain very good friends, and they serve as a reminder that no matter where you go in the Army relationships matter, and you're never truly alone. Hard work pays off. That season of challenge taught me about leadership, resilience, faith, and the power of community. I didn't just learn how the Army ran; I became someone others could depend on to help it run better. I was assigned to Headquarters, 1984th U.S. Army Hospital, Ft. Wainwright, Alaska. I had been told more than once, "This is probably going to be the coldest duty station in your career," and I was about to find out just how true that was.

My sons and I arrived in May 2000, landing in Fairbanks at 2:00 AM As we stepped off the plane, we expected the usual stillness of night but instead, we were greeted by a surreal orange glow across the sky. The sun had technically set, but the sunset glow hadn't gone away. It was one of those unforgettable Alaskan phenomena: the midnight sun. We didn't even have time to adjust before another surprise hit us: the mosquitoes. Huge, relentless, and everywhere. Locals jokingly called them "the state bird," and it didn't take long for us to understand why. On top of that, we were bundled up in jackets, bracing against the chilly air, while Alaskan natives strolled around in shorts, and T-shirts, like it was mid-summer in the lower 48 (the states below Alaska). That was our welcome to Alaska. It was two months before we received base housing and began to settle in. The boys were enrolled in school, and I started working on getting everything in place for our new life. As promised, the Command Sergeant Major's cousin who had been alerted of our arrival welcomed us like family. True to the Command Sergeant Major's words, she looked out for us in every way, helping us navigate life in this unique new environment. Her warmth and support made all the difference, and we remain close friends.

My responsibilities at the 1984th U.S. Army Hospital were like those I had held back in Maryland only now I was carrying them out in one of the most remote locations in the Army system. The unit operated from a large Reserve center located on post at Ft. Wainwright, and our mission was unique: we provided medical care and logistical support to Tripler Army Medical Center in Hawaii. That connection meant we were part of a broader network, extending our impact all the way across the Pacific. One of the unexpected blessings of this assignment was the opportunity to support Tripler, the pink hospital on the hill in Waikiki, directly. Our annual training was always scheduled in Hawaii, allowing us to not only work closely with our counterparts there but also to experience a part of the world most Soldiers only dream of visiting. Over the course of my three-year tour, I was fortunate enough to travel to Hawaii at least 15 times. Each trip offered a break from the bitter Alaskan cold, and a fresh reminder of the global reach of our mission. My son Stoney graduated high school and enrolled in the University Of Fairbanks. Not an easy choice, but I had to leave him to attend school and move to my next duty station.

Of course, the move to Alaska wasn't without its challenges, especially for my sons. The drastic change in climate, daylight hours, and distance from family took some getting used to. Winters were long and dark, summers were bright and short, only 98 days of summer, and the extreme dry cold was unlike anything we had ever experienced. The housing had to have a garage. Our vehicle had to have an electric starter, and a plug attached to the battery, because everywhere you went you had to plug up your vehicle, or it would not start when you returned. I remember the first time the temperature dropped below zero and I opened the garage to leave for work. That temperature device in my jeep was rolling so fast it looked like a slot machine. It finally stopped at negative 25 degrees below zero. I said *wow, welcome to Alaska*, but we adapted, just like military families always do. We all adjusted, made friends, embraced the adventure, got involved in one of the local churches, St. John Baptist, and learned how to thrive even when things weren't easy. Even in the harshest climates, I committed myself to staying positive, doing my best, leaning forward in my career, and continuing my education. I enrolled in Fairbanks University of Anchorage and continued my degree plan. Alaska demanded a lot, but I refused to let the cold or isolation slow me down. I stayed focused, receiving strong performance evaluations, continued my academic excellence, and taking full advantage of leadership training opportunities. Despite the freezing temperatures, I kept up with my physical fitness and maintained my health, knowing how important those habits were for my success and well-being.

One of the highlights of my tour was when my mother, Aunt Lillian, and cousin Janice came to visit. It gave us a chance to explore some of Alaska's most breathtaking landscapes together. We took trips down to Anchorage and Seward, witnessed the awe-inspiring glaciers, and marveled at the ocean's incredible sea life, a side of Alaska that few get to see. And then there were the northern lights, a spectacle that can only be fully appreciated in the extreme cold of an Alaskan winter. The skies would light up with colors so surreal, it felt like you were staring into heaven itself. And yes, contrary to popular belief, Alaska isn't dark year-round, the sun barely sets in the summer months, and the air is dry and fresh.

Like many duty assignments, this one came with its share of challenges. I faced some friction with the command team, moments that tested my patience and professionalism. But I kept my focus on the mission and my commitment to growth. The experience made me stronger, more resilient, and even more determined to lead by example. The time in Alaska was both demanding and rewarding. I learned how to serve in isolation, how to lead through extreme conditions, earned an associate of arts degree from the University of Anchorage, and navigated how to maintain focus on the mission while keeping my family grounded and strong. It stretched me professionally and personally, and it taught my children the value of resilience and the strength of unity.

By the end of my third year, my dedication had paid off. I was promoted to Sergeant First Class (SFC). With the new promotion came a new assignment and chapter. I received orders for my next assignment to the 865th Combat Support Hospital (CSH) Medical Unit in Utica, New York.

Leaving Alaska was bittersweet. Despite the challenges, it had become a place of growth and personal transformation, but I knew it was time to take on new responsibilities and continue developing as a leader.

The Sound of My Yes

The 865th Combat Support Hospital had a critical and clearly defined mission: to provide resuscitation, initial wound surgery, and post-operative treatment to injured personnel returning as many Soldiers to duty in the combat zone as possible. In peacetime, the unit operated under the 8th Medical Brigade, but when mobilized, it fell under a medical brigade or group that responded to global medical contingencies.

This assignment represented a shift. Not only was I back in the lower 48, but I was also stepping into a unit with a combat mission focus, which came with elevated expectations and an operational mindset. The team here trained with purpose, preparing for real-world medical support, under field and deployment conditions. I knew my leadership, logistics, and medical material management experience would be put to good use. As I settled in, I began to see how each assignment in my career had built upon the last. I was no longer just executing tasks, I was mentoring junior Soldiers, refining policy implementation, and shaping the overall readiness of the unit. This wasn't just about filling a position. It was about legacy, preparation, and service at a higher level.

The 865th Combat Support Hospital came with its own set of challenges not just professionally, but personally. From the moment I arrived, it was clear that this place would be different. There was a noticeable division in the surrounding community especially when it came to the school systems, housing, and daily interactions for people of color. It appeared to me that all black people were pushed to one side of town and had to deal with substandard schools and housing. It wasn't something you could ignore.

I wrote a letter to my career manager at Human Resource Command (HRC) and informed him of the situation and suggested they not send any more people of color to this location. First, it was felt in the small things: in the way people looked at you, and in the subtle but persistent barriers that existed in everyday life. Then, as I began to look for a place to live, and for my son to attend school, it was very evident to me. Despite the disconnect, I found a decent space to live in, and I made sure the school district was Blue Ribbon rated. I focused on the mission and my role and keeping my family safe. But my time at this organization was short. I reported in July of 2003, and three months later I was cross leveled into the higher headquarters to back fill a Medical Supply Specialist Military Occupation Specialty for a deployment mission.

The entire landscape of the military and the country shifted following the September 11, 2001 attacks, the deadliest terrorist attack in American history. Our nation entered what would become the Global War on Terror. Those attacks not only changed how Americans viewed safety, and service, but they reshaped our military mission entirely. The U.S. launched a multi-decade effort to dismantle terrorist networks, eliminate threats, and confront regimes believed to possess weapons of mass destruction. That era called on every unit to prepare differently, train harder, and move faster. We were no longer simply training for the possibility of war; we were now supporting an ongoing conflict, with real-time deployments, and high-tempo operations. As a result, my time at the 865th became a steppingstone toward something much larger. The momentum of that time would soon sweep me up into the next mission, one that would again test my leadership, resilience, and readiness in ways I couldn't yet imagine.

As the Global War on Terror intensified, 77th Army Reserve Command (77th ARCOM), the higher headquarters of the 8th Medical Brigade and the 865th Combat Support Hospital were preparing a small group of 32 personnel to deploy to relieve in place an active component medical command. The mission of the 8th Medical Brigade was to maintain the Theater of operations (military deployment into Southwest Asia) and ensure forward operations base medical facilities were equipped and prepared to take care of service members for sick call and battlefield injuries. If service members could be treated and returned to duty, it was done so expeditiously or they were moved forward to a hospital for critical care. I was not scheduled to be a part of that mission. The medical supply specialist (68J) Master Sergeant, and the command had a difference of opinion, and he was removed from mobilization. He and I were the same military occupation specialty, and three months after that incident, I received orders for deployment in support of Operation Iraqi Freedom (OIF). The mission was no longer theoretical. We were going into active combat zones, where lives were on the line, and every decision carried weight. The days of weekend drills and annual training exercises were behind me. Now, it was about real-world execution, logistical precision, and survival.

I reported to Ft. Wadsworth in New York for team briefs. We were scheduled to start Pre-deployment training at Ft. Dix, New Jersey in late November, early December. Because we had a general officer with us during pre-deployment training, we got pushed to the front of the whole training schedule. We had to complete four weeks of pre-deployment training that included everything we learned in basic combat training, low crawling in the mud, physical grass drill exercises, and obstacle courses.

Our unit mobilized quickly. Four days before we were wheels up (on the plane), my brother, a Marine, passed away from cancer. The Command did not want me to leave because that would delay the deployment and they would have to replace me in the ranks before the unit could leave. The unit would miss the window of travel already in place. There was an urgency in the air that we had never felt before. The Command Sergeant Major spoke with the Commander, and I was informed I could go home and see my family and brother, but I could not stay for his funeral service and if I did not return in three days, I would be considered Absent Without Official Leave (AWOL) which would end my military career. I pointed out that a few members were being left behind because they were missing some critical deployment equipment and once received, they would travel on a later date to meet with the staff in Kuwait. I was denied, because I was essential and I had to be a part of the main body. An executive officer escorted me to the airport in Philadelphia, and I was reminded that If I knew what was good for my career, I would return on the fourth day as the commander ordered.

I was filled with so much frustration, and many emotions, but the one that stood out the most was rage and anger. I told him I understood the order, but I was sure my original unit would contact me while I was home and let me know I could stay for my brother's services. I waited till the day before I was due to depart back to Ft. Dix for the Command to step in and allow me time with my family, and my brother, but that call never came. Filled with sorrow and disbelief, I stood strong for my sons and my parents. I wrote a letter of heartfelt expressions to be read at my brother's service in place of me not being present to give my own remarks. It began with, "I am unable to be with you today, because I am at War to keep America free." I was able to choose his suit for the service and view him with the family at the funeral home prior to me departing. I had to leave my sons, and family, and report to a war zone as my family celebrated my brother's life and laid him to his final resting place. When I returned, I broke down and cried. I was consoled by a few members of the staff, then briefed, and equipped for deployment into a region still reeling from the initial invasion.

I regret that to this day. I will never get that time, that moment back to say goodbye to my brother, and I still get angry today when I think about how I was treated, and no one came to my rescue to prevent that from happening. I live with that every day, an irreversible act. My assigned unit command team was not aware of the situation until I arrived in Kuwait, by then the deed was done. I had nothing for anyone the entire time I was deployed, I was angry when I left, and angrier when I returned. My team rallied around me and helped me keep my sanity. I am forever in their debt for all they did for me. I was determined not to let the command team leadership break me. I had to pray more often to calm my soul.

We arrived in the theater of operations in early January. We were quickly in processed and transported to Camp Arifjan (Southwest Asia Military Post). The departing unit Command team met us and escorted us to our living quarters, which was set up by rank and gender. Our staff was housed in tents for a few weeks until our barracks were assigned. We had to take showers and use the outdoor facilities that were created by the engineering units assigned to the past. The hygiene area was not near our tent. We had to walk in pairs to the area, because there had been several reported victims of rape on the post by this person called "Ghost." So safety was number one. Our general officer was assigned quarters on post, and a military aid to assist him with the mission. We all lived close to the hospital area of operations and the operations center because that is where we would perform our duties. We had about five days to learn our responsibilities.

The next morning, the departing team leadership picked us up from our tent area and began the right seat right (one on one training) for us to take over the mission. Each team member sat with their counterpart and was shown processes and procedures to continue the medical mission. I was trained by Sergeant Major Regina Rush Kittle on all her responsibilities. She took me under her wing and showed me things that I should know and places to get things to help me stay more comfortable. Before she departed, she gave me the rank of Sergeant Major to wear under my Battle Dressed Uniform shirt collar, and said "I am sure you will be a Sergeant Major one day." Even after she returned home, she sent me hair and hygiene products that worked well for her with the harsh base water and dry desert heat.

We eventually were assigned to a large bay in the barracks across post, and the bathroom was right outside our area. Thank God the person they called "Ghost" that committed those sexual assaults was caught, legally charged, and removed from the military, but I remained vigilant and never went anywhere alone. The deployment was long, and many things happened that were unfair, and unjust to me and other members of the unit. We were all requested for our Military Occupation Specialty skill set, so each of us had a key role in the deployment area of operations. The General Officer in charge didn't see it that way. He chose whom he wanted to be a member of his team and the rest of us were assigned to a makeshift unit called the Provisional Battalion. No such structure existed in the U.S. Army. We looked around and the unit was mostly comprised of black and brown men, and women, while the command staff was all white. Major Al Rogers (Uncle Al) was also assigned to the makeshift battalion as the executive officer. He pulled us together, gave us guidance and protected us from all the mayhem from the crazy missions, and assignments the command staff would task our team. We became the Head Quarters staff that handled all the daily duties. We would perform daily updates of our mission in the morning and afternoons to provide a battle update, and when we finished briefing, we had to depart the area. We felt like misfits, but we all performed like champions. We were responsible for all the medical treatment facilities throughout the Kuwait area camps. We would travel from post-to-post ensuring teams were well staffed, trained, and equipped. I got a chance to work with the Navy combat support hospital from Portsmouth, Virgina, a Navy unit that oversaw the mobile trauma hospital on base. That was a phenomenal experience.

The heat was relentless, and the environment was unlike anything I had experienced before. We traveled everywhere on post by foot, no matter the temperature. The dining facilities (mess hall) were open for meals four times a day, because of shift work and travel teams arriving late in the evening. The temperature during the day would reach 150 degrees and there were pallets of water everywhere. The dining facilities would have large refrigerators full of every beverage except alcohol. Kuwait was considered a dry country, but there was alcohol available if you had the right connections with certain civilian contractors. The Mess teams would prepare special meals like Surf and Turf, Mongolian stir-fry nights, rich creamy ice cream, and all sorts of desserts. The work environment was in a state of turmoil. No day passed without the thought of *today may be my last*. We would travel to Navy vessel and supply warehouses to receive military and commercial vehicles and supplies. We traveled in areas that had no roads, experienced severe blackout sandstorms, extremely long convoys, the constant sound of helicopters overhead, and guarded gate entrances and exits. Security check points and bus traffic jams all became part of the daily routine. But nothing ever truly felt routine. You stayed alert. You stayed sharp. You worked 10-16 hours a day. You kept your team close.

Every patient that came through our tents or base operations reminded us of the importance of our work. These were sons, daughters, brothers, sisters, real people, real stories, real sacrifice. I took pride in ensuring our medical logistics flowed efficiently. Supplies arrived on time, equipment was tracked and maintained, and the medical staff had what they needed to perform their jobs under the most stressful conditions imaginable. I learned quickly that leadership in combat wasn't about rank, it was about reliability, integrity, and trust. You had to be calm in chaos. And sometimes, you had to lead without being asked. I kept busy in the desert. I was up every morning running at 4:00AM with Soldiers during physical fitness exercise and staying fit. I lost 20 pounds sweating in that heat. We ran five miles in the early mornings, or late into the evening to avoid the heat.

I enrolled in college courses and took classes twice a week, attended weekly Bible study and was a member of the Camp Arifjan choir. The choir was a combination of individuals, units and armed forces. We practiced at the church temple twice a week and sang during church services on Sundays. We had so many talents, musicians, soloists and praise dancers that made our choir unique. We would pack the main chapel every Sunday, service members and military contractors would be sitting in the aisles, on the steps and standing along the walls of the chapel to hear us perform and receive the spoken word. We were asked to perform at several post events in the desert to build and sustain morale and remind them of God's promises. I still have my song sheets from practice in a binder with my Army keepsakes. I participated in the post Soldier morale events often; one was a modeling show to help build confidence and self-esteem. Clothes were donated for us to walk the runway and strut our stuff. We made the best of our time there and I am forever grateful for the love I was shown by my team and choir. It was because of them that I was able to hunt the good stuff and keep my focus on praise and worship.

I earned two coins of excellence while serving there, one during the annual physical fitness test, because I outscored the command team with 400 points using the extended scale of measure. My anger was still very present almost a year later, and I know it was the driving force to beat them. The second coin was awarded by the Command post Command Sergeant Major, for having the cleanest area and bathroom in my living quarters on the entire campus.

The time passed and our staff was preparing to return home. We received our re-deployment orders and followed the same process as the unit before us to prepare for the incoming unit staff. We packed up our Conexes with supplies on the last day in the theater; there had been whispers of units getting rerouted to new missions that had to report immediately. We were ready to go and crossed our fingers that we were not one of those units. We proceeded to load our personal gear onto the bus and headed inside the auditorium for the change of responsibility ceremony.

Once the ceremony was completed, we were heading to the bus when a team of Soldiers entered the auditorium and blocked the doors. Right away my thoughts were, *OMG, we are being rerouted*, but that was far from the reason why they were there. An officer made his way to the stage in the auditorium and began speaking on the microphone. He stated, "We are the Command Inspector General team, and you are being delayed to provide information on a formal investigation, if I call your name, report to the stage. Everyone else can leave."

Several team members were called to the stage, including me. We were asked several questions: each of us had to write and sign a sworn statement, then we were released to board the bus. During my time there I became aware, both directly and indirectly, of situations where individuals engaged in unprofessional relationships and encountered unwanted sexual advances. Those actions not only compromised the values we strive to uphold, but also created an environment of mistrust, resentment, and inequality among those who carried their responsibilities with integrity. While I've always believed in mentorship, fairness, and hard work, it was difficult to watch the damage these choices caused not just to morale, but to the reputation of those of us who worked honorably and led with excellence.

When we loaded the bus, we noticed several members were detained, including our General Officer, who was further detained and not allowed to depart with us. No one spoke a word until hours later when we landed at our first re-fuel point. Then it was disclosed that team members had placed a few several-page formal complaints against our Command staff about how we were treated, and the inappropriate behavior between certain staff members during the deployment. Months later I learned that some formal charges were administered, and legal actions were taken against them. This deployment shaped me; I learned leadership is not about barking orders, it's about serving your people, making decisions under pressure, and finding strength when fear whispers in your ear. We built a family in the desert, and those bonds will never break.

When I returned from deployment, one officer truly stood with me during the most difficult season of my life, Captain Tianika Mangum, now Lieutenant Colonel Mangum. She supported me as I faced the loss of my brother and endured the pain of those who did not stand with me in my grief. Through her steady presence, a genuine bond of trust and friendship was formed, rooted in shared faith and quiet strength. She kept me grounded and focused on healing when I needed it most. That debt of gratitude can never be repaid, and for that, I am forever thankful.

After completing my deployment, I returned with a deeper sense of purpose, hardened experience, and the weight of having served during a pivotal moment in history. I had grown not just as a Noncommissioned Officer, but as a leader, mentor, and Soldier who had seen firsthand what it meant to serve in the most chaotic and uncertain of times.

"God is your strength."

The Sound of My Yes

Crossing the Threshold
The hero leaves her comfort zone.

CHAPTER 5

FROM ORDERS TO OUTCOME

"For I know the plans I have for you."
Jeremiah 29:11 (NKJ)

Not long after returning, I received orders for my next assignment: Army Reserve Medical Command (AR-MEDCOM), located in Pinellas Park, Florida. My sponsor Evelyn Bonner (Felton) contacted me prior to my arrival and provided all the information I needed for getting settled in Florida. She was, and still is, a lifeline for me, then and now. I knew this assignment would mark a shift from the operational tempo of a combat support hospital to a more strategic and administrative environment. But make no mistake, the responsibilities were just as critical.

Army Reserve Medical Command served as the central command for all Army Reserve medical units, overseeing readiness, training, mobilization, and support for global missions. It was a place where big decisions were made, where policies took shape, and where the future of Army Reserve medical support was guided. To be assigned there meant I had reached a level of trust and experience that positioned me to help influence the broader mission. My experience in logistics, personnel accountability, and command support made me a natural fit for the work.

I was assigned to support high-level operational planning, medical readiness reviews, and even special projects that helped prepare units for future mobilizations. I had a chance to contribute to the kind of strategic planning that affects units across the globe. Army Reserve Medical Command expanded my role in ways that tested and strengthened my leadership at a new level. I was no longer in the trenches of day-to-day logistics, or combat medical support. Now, I was helping shape the strategic direction of Army Reserve medical readiness across the country.

I was assigned to a team of eight team members with different Military Occupation Specialty. We were responsible for conducting readiness inspections of all the units within our area of responsibility. Our official duties included coordinating personnel movements, supporting medical readiness reviews, and ensuring that subordinate units were properly resourced and trained for mobilization and deployment. I also had the opportunity to work closely with officers and senior enlisted leaders across multiple commands. My combat experience gave me a strong foundation, and I quickly became a go-to voice for practical insight on how policy impacted boots-on-ground Soldiers.

Another leadership milestone came when I was selected to facilitate planning conferences that prepared subordinate medical units for readiness exercises, and deployments. I helped develop training timelines, ensured commanders readiness reports were accurate, and assisted in standardizing processes that impacted thousands of Soldiers. I also pushed myself to grow professionally. I enrolled in advanced leadership courses, like First Sergeant course, and Battle Staff. I pursued higher education at least once a year and stayed active in the community. I became intentional about preparing myself not just for the next rank, but for life after the uniform.

That forward-thinking mindset was something I began sharing with others too. I encouraged Soldiers to think beyond today's

From Orders to Outcome

mission: to plan early, stay flexible, and build a life with purpose. This assignment taught me that transformational leadership doesn't just happen in combat, it happens in conference rooms, one-on-one mentoring sessions, and even casual conversations in the hallway. I began to truly understand the ripple effect of strong leadership, and how investing in people early and often was the most impactful thing I could do. My time at the Army Reserve Medical Command reinforced a truth I had come to live by: leadership isn't about the position, it's about the influence. And that includes the influence you have when you choose to serve others, guide them, and equip them for the road ahead that stays with them forever.

One of the most rewarding aspects of my time at the Army Reserve Medical Command was meeting the man God had prepared for me. When my mother visited me in Florida in October 2005, the year my dad passed, she said, "Pammie, I'm going to pray that God sends you an Angel to take of you and your sons, like your father took care of me and all our children."

Two months later, I met Calvin randomly at work while looking for another coworker. When I arrived in the work area of my coworker, Calvin was sitting in the same cubical workspace. I asked him if he knew when the person I was looking for would return, and he responded, "No," and turned back toward his computer. I then asked, "Are you new to the Command?" and he replied, "Yes. I arrived with the mobilization group." I introduced myself and invited him to meet my group of friends, Senior Noncommissioned Officers. Later, I learned Calvin was a part of the reserve Soldier Mobilization that was brought on active duty from North Carolina to support the mission of Army Reserve Medical Command.

Later in the month, Calvin joined our table during the annual Christmas party celebration luncheon the command hosted every year. I introduced him, and we all sat and enjoyed the time together. While we were cleaning up from the party, I invited him to join us for a staff Christmas party to meet other people at the Command. He said maybe, so I gave him the date, time, location and address, but he never showed up. After the holiday break, I saw him in the employee parking lot and asked him why he didn't come to the party. He apologized and said he decided to spend time with his family for the holidays. I told him since he did not show up, he owed me lunch. To my surprise, he agreed and set a date and time for later in the week.

We were both new to the area, and I really didn't know where any good restaurants were located. One of my team members recommended a local deli called Roe's, so I made reservations. We met at the restaurant and chose an area to sit. We began small talk and connected instantly. That day marked the beginning of something extraordinary. The relationship unfolded quietly. During our meal, I remember asking him why he was interested in me. He chuckled and said, "It was the velvet skirt you wore to the company Christmas party. But more than that, it was your language." It was how interested and eager I appeared to share conversation that intrigued him.

Both of us carried the weight of past failed relationships, so we were very cautious about rushing into anything new. We decided to take it slow, but from that first lunch, our nightly conversations grew longer and more meaningful. We lived about 45 minutes from each other, but we would spend hours laughing, sharing our hopes and dreams, and imagining what life might look like with a soulmate.

The weekend after our lunch, we attended a friend's birthday party, not as a couple, but as two friends simply enjoying the evening. Neither of us revealed our budding relationship, unsure if we were truly a match. Still, we continued to see each other every day, talking endlessly, building a connection that felt both natural and undeniable. We simply were enjoying each other's company. What truly touched me was his perspective on the past. He told me, "What is in the past is done, and today we start fresh." It was a simple yet profound statement that convinced me to trust him and continue to build this relationship. I found myself acting like a high school girl. I created a scrap book of all our first activities together, like going to the movies for the first time. We had a friend that was into photography. He would meet us in different areas like the park and church events and take pictures.

After a few weeks, Calvin asked if I would like to exclusively date him. I paused, unsure what he meant at first, and he clarified only me, and only him. I laughed, thinking that was the obvious way to date, but he was serious. He wanted clarity and commitment. I agreed, and with that understanding, we began sharing more of our past experiences and future desires. I shared my own past, failed marriages, shattered relationships, and my fears rooted in past betrayals and lies. Trust and loyalty were non-negotiable for me. I told him I could handle the truth. I was independent, having secured a comfortable life for myself and my sons, and I didn't need financial security from a partner. What I wanted was love, companionship, and a partner who valued commitment, integrity, faith, and loyalty as much as I did.

Within just two months of us dating, unknown to me, Calvin purchased a custom-designed one-carat solitaire engagement ring from a well-known jewelry store, Mayors. We celebrated his birthday in February and Valentines Day was just a few days after. We had been working late all week with several unit reports, but we found the time on Valentines Day to have a nice dinner.

I expected plans to do something on Valentines Day, but I did not expect the gift I received the day after Valentines Day. He called me late that evening and asked if he could come over to my house. I asked if he wanted to make the drive that late in the evening. He replied, "I know its late, but yes, I want to see you." He arrived and came into the living room where I was watching television, and started small talk and then he got down on one knee and said, "I know we have only known each other for a short time," he pulled the ring out of his pocket and said, "but I know I love you and I wanted to ask you to marry me." I was speechless. I said, "Are you sure you are ready for marriage?" He responded, "Yes," but he wanted to date for a year first, to see if we truly were equally yoked. I accepted his proposal, and waiting a year was not a problem, because I had said yes too soon a few times before, and I needed to be sure this time. The ring was a little big, but of course I wore it.

Once I arrived at work, I didn't say anything, but it was humorous watching the reactions. My friends were hitting me left and right with questions, concerns, and recommendations, but over time, they settled down when I shared that we were waiting a year to be sure of our commitment. A few days later, we went to the jeweler to get the ring sized. She smiled and greeted us, then remarked how rare it was to see someone pay cash for a ring.

Just three months later, Calvin wanted to upgrade my ring to a larger diamond. He then upgraded to a 1.50 ctw solitaire diamond. Usually, it is the women who asks for an upgrade. I was overwhelmed by his love and generosity, and my family adored him from the start. We had an amazing courtship. We did so many things together, conversations were pleasant and pleasing, no ambiguity or arguments, our families were pleased and accepting. Loving Calvin was easy. And that was when I knew my mother was right: he was my angel created by God just for me. When things come easy and naturally, and there is no wavering or stress. I knew this union was of God.

We have been married 19 years now, and after the second year of marriage, he upgraded my ring again to a beautiful pear shaped 2.50 ctw carat solitaire and he added an anniversary all diamond band. This time he said with a smile, "This is the last upgrade," and the truth be told, I loved all of the rings. After all, it is not the ring itself, it is the man behind the ring that counts the most. Reflecting, I realize I look for the qualities in a man that mirrored my father's, someone who would provide, protect, and show unconditional love. I see now that I married a man just like my dad. He shares his kindness in small ways like buying donuts, doing the grocery shopping, keeping a little "Johnson store" stocked with essentials, just like my father did for us. His actions remind me that love, loyalty, and stability are what truly matter, and with Calvin, I found all of that and more.

Me and Calvin on our wedding day.

All the relationships and friendships developed during that time at Army Reserve Medical Command were special. We all became a family and twenty years later we are still connected and remain very close. During my tour at Army Reserve Medical Command, I served in many roles as the master fitness instructor, and my all-time favorite duty was the Honor Guard Noncommission Officer in Charge (NCOIC). I established the Honor Guard team and would conduct physical fitness every Monday, Wednesday, and

Friday. Those were some good times doing circular training and group led runs around the command area. Those programs improved health and wellness for everyone that participated, and I was able to continue my 300 plus perfect scoring streak on the physical fitness test and earn a few more fitness badge patches.

There were 12 Honor Guard team members. We were trained by the Human Resource Command Honor Guard team, which received their training from the Old Guard in Arlington, Virginia. We would spend hours practicing with flags, weapons, and a saber (Noncommissioned Officer sword) in our battle dressed uniform (dress blue hats, with the wide patten leather brim, placed just over our eyes, white gloves and black dress shoes, with enforced leather soles, and steal taps on the front, back and side). We would perform honors at many events; the team was sharp. If you google Army Reserve Medical Command Color Guard, a video will show our team in action. This was a very proud time for me; I was happy to be part of this organization. The team remains intact today, still rendering honors across the Florida area. I kept in contact with several members of the team even when I departed.

Six years after I left Army Reserve Medical Command, I received word that One of my Color Guard members Master Sergeant Steven Keith Graham (age 38), whom I called my son, was in an Atlanta, Georgia hospital in critical condition. I rushed at that same moment to book a flight early the next morning, but he passed away before I could arrive. I still hear the phone ringing in my head every time I think of him. To this day, I have not brought myself to delete his number from my phone. He was a great Soldier, father, and friend, and the Army Reserve Medical Command Tampa family misses him dearly.

I served three years at the Army Reserve Medical Command. During my last year, I received notice that I had made the selection list to attend the most prestigious professional military education opportunity available to a senior enlisted leader, attendance at the United States Army Sergeants Major Academy (USASMA) Ft. Bliss Texas. There are two pathways to attend USASMA Resident and Nonresident both structured as 12-month programs. This course was the Senior leader course required to make Sergeant Major. It was the next step in my leadership journey, and a reflection of everything I had poured into my career thus far. The weight of it didn't just rest in the rank, it resonated in my soul. It meant more than career advancement. It meant we, women of color, were in the room. Being a Black woman selected for the Academy is not just about achieving a milestone in military service, it's about navigating a terrain that was never built for us and choosing not only to walk it but to rise on it.

The truth is, being selected was only the beginning. The Academy is where the Army sharpens its top enlisted leaders, where doctrine meets decision, and where leadership becomes legacy. But for a Black woman, it's also where silence meets strength, where you learn to navigate the unspoken tension in classrooms where few look like you, where you work twice as hard, not because you're trying to prove yourself, but because you know that when you speak you don't just represent your experience, you carry a lineage of people who were often overlooked or underestimated. From a group not always expected to succeed at the highest levels, our class commissioned over 40 Female Black Sergeants Major into senior enlisted leadership.

I was initially selected to attend the USASMA Nonresident Course, Class 36. The program was structured in two phases: Phase I consisted of distance learning, followed by Phase II, which included two weeks of resident instruction at the academy. The 12-month course required me to maintain my full-time duties while completing academy coursework during evening hours. A week later a peer who was on the same list was selected to attend the resident course, which was nine months at Ft. Bliss, El Paso, Texas school house, contacted me and asked me to switch class assignment with him because he was scheduled for a deployment. I accepted and received orders to report in August 2009.

Class 60 was unlike any class that had come before. The course was newly revamped, with a comprehensive curriculum modeled after accredited college courses, bridging the gap in knowledge between Senior Noncommissioned Officers, and Senior Commissioned Officers. The class attendance was over 400 master sergeants, and a few reserve Sergeants Major, some foreign senior enlisted students, and joint service members from the Navy, Marines and Coast Guard. It was the first class that had over 48 women of color in attendance and to graduate. This wasn't just about military doctrine anymore, it was about strategic leadership, critical thinking, organizational development, and the evolution of our Army in a modern world.

The course demanded everything we had in mental agility, emotional resilience, and strategic depth. We engaged in discussions that challenged our leadership styles and forced us to think beyond just unit-level operations. We covered subjects like joint force integration, ethical leadership, global operations, and strategic communication. The expectation was that every graduate would walk away prepared not only to wear the rank but to embody the vision and voice of the enlisted corps at the highest level. This was the final senior leader course for promotion to Sergeant Major. It was understood that completion of the course guaranteed promotion within two years.

I got settled in my apartment and asked around about a church home to serve in while I was there. I was told to visit St. John Missionary Baptist Church if I wanted some good Bible-based ministry. It was a small church led by Pastor Michael Grady and the first lady, Jenevelyn Grady. It never ceases to amaze me how God will guide you where you need to be. It was the best decision I made to serve there. The services and Bible studies were always very personable and meaningful. It really felt like my home

From Orders to Outcome

church. Pastor Grady, who I call my adopted father, and I were from the same hometown. He is truly an anointed man of God. He started his own house of worship, Prince of Peace in El Paso. We remain in contact, and I know I am blessed and a better vessel for God because our paths crossed.

Because there were over 400 students, the classes were split, half would attend morning hours 9:00 AM to 12:00 PM and the other half from 1:00 PM to 4:00 PM. I was assigned to the morning classes. About a month into the course, while in class, I received a notice via email that I had been selected for Sergeant Major course to attend residence course for Class 61. It's an unusual honor to be selected twice for the same opportunity, especially when it happened because my board file was inadvertently submitted twice.

While the duplication wasn't intentional, I'm grateful the selection board saw value in my record both times. It's a reminder to stay ready, stay humble, and trust that what's meant for you is for you. The commandant of United States presented me with a congratulations note on my selection for the course, and making the Sergeant Major selection list. I was confused. I was already in the Academy attending Class 60. I contacted my career advisor to determine what happened and was informed that my board file was submitted and reviewed on two separate occasions, resulting in being selected twice for the same course. This created an unexpected debacle, prompting questions about process integrity and fairness not just for myself, but for others competing for the same opportunity.

While I was honored by the selection, I recognized the importance of transparency and accountability in board procedures and remained committed to resolving the matter with professionalism and integrity. This was ultimately resolved, and I continued in Class 60.

During the last three weeks of the course, our assignment was a large battlefield operation that encompassed full scale training exercises of war time simulation. Every student was assigned a specific task, and we had to perform as if we were at war. One day while in class I received an email asking me to report to the Active Guard Reserve Liaison office on campus. I replied, acknowledged, and when released for class break, I reported it to the office. I stated, "Sergeant Major, you asked to see me?" He didn't say a word; he pushed a piece of paper across his desk toward me to pick up and read. At this time in the course, everyone was receiving their follow-on assignment orders. My assumption was, he had received my next assignment order.

I was completely surprised at the content on the paper. It read "The United States Army has proposed special trust in your leadership, and you are hereby promoted to the rank of Sergeant Major, effective 1 April 2010." I thought it was an April fool's joke, I asked him if this was for real. "Is this a promotion order for Sergeant Major?" He simply said, "Congratulations." I was in shock, another rare and humbling blessing. I returned to my class and did not share the news with anyone. I called my husband, Calvin, and then my mom, to share the good news.

Class 60 was unique not only because we were the first class to study a more inclusive curriculum and bridge the gap of knowledge between Senior Noncommissioned Officers and Officers in the command. It was also the only class that every active component Master Sergeant that passed the course would be frocked (pinned without pay) to Sergeant Major, with a promise of within two years would be guaranteed promotion to Sergeant Major. I did not want to blast this news, and make peers feel some kind of way. I had a small ceremony during the main exercise and showed up for the afternoon classes with Sergeant Major rank on my uniform. I answered the question as peers noticed the change. Eventually, I had a party to celebrate my accomplishment. With God's hand in motion, I was the only student in Class 60 to be promoted to Sergeant Major with full pay two weeks before graduation.

I'll never forget that moment standing tall at Ft. Bliss, wearing the rank and receiving Sergeant Major pay, a symbol not just of promotion but of purpose fulfilled. I walked those Academy halls not as a guest, but as a rightful heir. And yet, I still had to bring all of me into a space that wasn't always designed to see all of me. I had to sit in lectures on leadership while my mind quietly negotiated the nuances of leading while Black, while female, while constantly being observed, but not always heard. I had to push past moments of isolation with purpose. I had to pour into younger Black women in the ranks behind me to let them know, "You belong here too." Years of sacrifice, deployments, cold nights in Alaska, sandstorms in Iraq, long hours behind desks, mentorship in the hallways, and prayers whispered in private all came together in that promotion.

But this wasn't just a personal win. I carried the names and faces of every Soldier I had mentored, every leader who had shaped me, and every family member who had supported me. My mother's and my aunt's pride, my children's resilience, my mentors' belief in me, they were all part of that moment. The Academy not only validated my experience, but it also gave me a refined voice as a senior leader. I left fully equipped and fully promoted, ready to take on my next challenge with a broader perspective, deeper conviction, and a renewed commitment to serve, lead, and inspire.

Fresh off graduation from the United States Army Sergeants Major Academy, I reported to my next assignment as the Operations Sergeant Major for the 330th Medical Brigade, located at Ft. Sheridan, Illinois. This position wasn't just another stop in my military journey, it was a chance to fully walk in the rank of Sergeant Major, bringing with me every hard-earned lesson, every challenge overcome, and a clear mission: to lead with purpose and mentor with intention.

The 330th Medical Brigade was a command with vast responsibilities. We were responsible for supporting and preparing multiple subordinate medical units for mobilization, training, and deployment readiness. As the Operations Sergeant Major, I sat at the heart of all planning efforts coordinating training plans, overseeing readiness reporting, and ensuring that the Brigade remained fully mission capable. But more than the operations and logistics, I saw this assignment as an opportunity to invest in people. I had climbed every rung of the enlisted ladder, seen both the battlefield and the boardroom, and now it was time to reach back and help raise up the next generation of leaders. I made it a point to sit down with young Noncommissioned Officers, especially those new to the Active Guard Reserve program or just beginning their careers in leadership. I shared my journey not just the accolades, but the trials, the detours, and the faith that carried me through. I taught them about accountability, strategic thinking, and the importance of maintaining their physical, mental, and spiritual well-being.

One initiative I'm most proud of was helping develop a mentorship program within the Brigade, pairing junior leaders with seasoned Noncommissioned Officers. Recognizing that too many Soldiers were being failed by the system not by lack of effort. I developed a Physical and Health Fitness Boot Camp for those struggling to pass the Army physical fitness test. I partnered with the Great Lakes Navy Base Health and Wellness Center to uncover the real reasons behind their challenges, providing access to X-rays, lab work, and nutrition support resources rarely available to Reserve Soldiers during standard drill weekends or annual training. With funding secured from higher headquarters, I assembled a Team Captain Staff Sergeant Leon Mangum who had the same passion and fire as I did and a team of Master Fitness Trainers and high-performing Reserve Soldiers to lead the effort. The results were undeniable. Soldiers passed, confidence was restored, and careers were saved. More than anything, they were grateful that a senior leader took the time to look beyond the scorecard and get to the root of the problem. We focused on everything from career development and promotion readiness to balancing health. family, education, and service

We focused on everything from career development and promotion readiness to balancing family, education, and service. I didn't want them to just survive the system, I wanted them to understand it, navigate it, and thrive in it. Leading at this level came with new challenges. Nothing meaningful is accomplished alone, and my success was rooted in the strength of the team around me. I had the privilege of working alongside an exceptional group of dedicated professionals, without them, our mission tasks would not have been accomplished. I owe a deep debt of gratitude to Lieutenant Colonel Brian K. Jones (Ret.), who never hesitated to tackle the hardest asks and consistently made things happen we once thought were impossible, and to Master Sergeant Jamie Jacobs (Ret.), whose shared knowledge, steady presence, and unwavering support became my solid rock as I learned to navigate and succeed in my role as an Operations Sergeant Major.

I dealt with policy implementation, engaged with senior officers, and sat at the table for strategic decisions that impacted the lives of Soldiers and their families. I learned how to speak truth to power, when to fight for resources, and how to protect the integrity of the enlisted voice in rooms where it often went unheard. Even more importantly, I continued to be a bridge between the junior and the senior, the strategic and the operational, the vision and the execution. Being at Ft. Sheridan allowed me to travel extensively, complete my college Bachelor's degree in Workforce Education Cum Laude with 3.8 GPA, engage with unit leaders across the region, and advocate for Soldiers in ways I never imagined early in my career. I regularly reminded myself and others: "This uniform is temporary, but the impact we leave on people is permanent."

Looking back, this assignment marked a powerful transition in my leadership. I was no longer just climbing the ladder; I was holding it steady for others. And as I poured into them, they poured right back into the mission, ready to carry the mantle of leadership forward. After completing my assignment at the 330th Medical Brigade in August of 2014, I reported to 3rd Medical Command (Deployment Support) (MDSC) at Ft. Gillem, Georgia, as the Chief Operations Sergeant Major. This was more than a new duty station; it was a leap into a role where global decisions were being made, and I was at the center of the planning table.

The 3rd Medical Command (Deployment Support) is a division-level command, and our team supports critical operations across the globe. From planning missions in Qatar, to assembling and deploying the first Army Reserve Medical Augmentation Team in support of 8th Army G-4 (Division level for logistics) Key Resolve training exercise in Korea, we were shaping the Army Reserve's footprint in real time. Our successful execution not only expanded training opportunities in the Pacific Command (PACOM) theater, but it also reinforced our presence and partnerships across allied nations.

As a medical planner for U.S. Army Africa and the 101st Sustainment Brigade, I helped shape the crisis response to the Ebola outbreak in West Africa. I participated in Joint Operations Planning and Execution System development for the Emergency Ebola Operations Center, ensuring uninterrupted assistance that contributed to halting a global epidemic. It was a heavy responsibility, but an honor. I also developed a G3 (division level of Army Operations) Mission Readiness Tool that improved data flow for Operation Shamrock, ensuring unit readiness was visible, timely, and accurate. My contributions extended into Operation Life liner support for Senegal and Liberia, uniting logistical coordination with humanitarian response.

While I was leading from the front and executing these complex missions, my personal life and spiritual growth were being shaped in profound ways. Being stationed in Georgia brought me closer to extended family, and I was able to reconnect with roots that grounded me during the fast pace of senior leadership. But the demands of the job were real. Countless late nights, never-ending contingency briefs, and endless coordination across time zones—these things took a toll. Yet, in the middle of it all, I was reminded of a deeper truth. I often paused to reflect on how God placed people in my path—mentors, battle buddies, leaders, and family—each of them playing a role in helping me stand in my purpose. It was God who went before me, opened doors I didn't even know existed, and kept a shield of protection around me in my most vulnerable moments.

The work I did at 3rd Medical Command (Deployment Support) was strategic, global, and intense, but the lessons were deeply personal. I enjoyed my time serving on the mission that made a global impact. I had to remember to stay ready because opportunity and challenge rarely announce themselves, stay faithful because what you can't see God already has covered, stay humble because the mission is bigger than the rank, position, or any accolades received.

Once I completed my assignment at the 3rd Medical Command (Deployment Support), I transitioned into one of the most dynamic roles of my career at United States Central Command, a four-star command based at MacDill Air Force Base in Tampa, Florida. This was a pivotal assignment, not just in terms of responsibility, but in how it broadened my scope of influence and deepened my calling to serve both nation and people through mentorship, diplomacy, and faith.

Initially, I was selected to serve as the Senior Leader on the Army Reserve Engagement Team within the G1 (Division level for Personnel) Division. This assignment required a reassignment into one of my original Military Occupation Specialties, 42A (Human Resource Specialist) and pulling once again from that toolbox of experience and networks built over decades.

My mission was to ensure that all joint service members across U.S. Central Command were deployment-ready physically, administratively, and emotionally while also ensuring their families were taken care of during their service. A few months into the assignment, the Deputy Chief of Staff personally requested I take on a much broader role: Headquarters (HQ) Commandant Senior Enlisted Leader (SEL), mentoring and guiding a diverse multi-service staff in support of split-base operations. I was responsible for supporting over 4,000 joint personnel: Soldiers, Sailors, Airmen, Marines, and civilians at headquarters, and an additional 400 personnel stationed across 20 countries in the Central Command Area of Responsibility. This role demanded 360-degree leadership: managing base operations, ensuring logistical and administrative support, and often serving in the absence of the Command Sergeant Major to address all enlisted matters. It was humbling and intense, but each challenge strengthened my belief that leadership is about service—consistent, compassionate, mission-focused service.

After a year at HQ, I was handpicked and reassigned within the command to the Strategies, Plans, and Policy Division J-5. The J-5 (Joint military service, five-identifier for Strategic plans) mission is strategic advising the Chairman of the Joint Chiefs of Staff on military strategies and policy recommendations aligned with national security interests. As Senior Leader advisor for security cooperation support, I worked directly with the Office of Military Cooperation at the U.S. Embassy in Uzbekistan, coordinating key leader engagements, and partnering with the Uzbekistan Ministry of Defense Special Operations Forces. We helped advance critical national security objectives while building enduring relationships with foreign military leaders.

One of my most fulfilling contributions was helping to develop a Women's Noncommissioned Officer Development Program with the Jordanian Armed Forces, designed to empower women in uniform and elevate their contributions through structured leadership education. Our goal was to establish a shared vision for Noncommissioned Corps development, enabling long-term capacity building within the Jordanian military.

This assignment revealed how truly global the reach of leadership can be, and how mentorship isn't limited to your branch, country, or language. It also reaffirmed something I have always believed. I know now, more than ever, that God directed my path, went before me, paved the way, and opened doors I couldn't even see. He shielded me in my every time of need, and put great people in my path, like Major Quana Wright, Chief Brenda Myers and Colonel Sandy Sadler. Quana and Brenda (my sisters and BFFs) helped me understand the Joint environment, and to this day keep me grounded.

My work at Central Command was a high point professionally, but also spiritually. It was here that I truly began to understand my post-military calling: to mentor, guide, and serve the next generation of leaders, whether in uniform, in schools, or in boardrooms. I was finally a voice for others.

My final assignment, not by choice but by chance, was at the United States Army Human Resource Command, Ft. Knox, Kentucky. This was a capstone role that allowed me to serve at the strategic epicenter of the Army's career management enterprise. The Human Resource Command complex was named and dedicated in honor of Lieutenant General Timothy J. Maude, who perished on September 11, 2001 in the attack on the Pentagon. At his time of death, was serving as the United States Army Deputy Chief of Staff for Personnel, G-1. The complex is the largest single building project in the history of Ft. Knox. It is a three-story, six-winged, red-brick facility. Human Resource Command is a direct reporting unit supervised by the Office of the Deputy Chief of Staff for personnel, G-1, focused on improving the career management potential of Army Soldiers. Human Resource Command also supports the United States Army National Guard, Army Reserve in their management of the Selected Reserve. The Human Resource Command Commander is also the commander of the Individual Ready Reserve, the Standby Reserve, and the Retired Reserve.

I had planned to retire in 2017 and start the next phase of my life. Usually when you elect to retire after 20 plus years of service, you get a one- to two-year transition period. The Active Component (AC) and the Active Guard Reserve must provide different documents in the request to retire. Active Guard Reserve Soldiers must provide every order for 20 plus years of service to validate federal service, where Active Component time is continuous, so orders are not required. During the two years, you complete and submit for approval a retirement packet, that consists of a memorandum requesting your date of retirement, permission to participate in an internship or fellowship with civilian companies for potential employment and all Defense Agency (DA 31) for terminal leave (last days of military Duty), participate in a transition assistance program (TAP) that provides information, tools and training to help service members and their spouses get ready to successfully move from the military to civilian life.

From start to finish, Transition Assistance Program guides you on veteran benefits, education options, federal assistance, and veteran employment help. Think of Transition Assistance Program as a lesson in understanding how to shift your life from being centered on the military to being focused on the civilian world. I learned of my reassignment in December 2016, while serving at the U.S. Embassy in Uzbekistan. I did what any committed professional would do: I sought counsel. I spoke with my career advisor to express my hesitation and concern about the fit and timing of the assignment. After a thoughtful conversation, I was presented with two options: report to the new duty station or retire within 60 days. I would just be arriving back in Florida in 60 days; it was an impossible choice. It wasn't an easy decision to shift my thought process from the idea of retiring to continuing to serve.

The unexpected nature of the reassignment forced me to reflect deeply on my values, my goals, and what I still had left to give in uniform. Ultimately, I chose to report, not out of obligation, but out of faith that there was still purpose in my path, even if I couldn't see it clearly at the time. I received orders in January to report in May 2017. A few weeks later I received a very nice letter from Human Resource Command welcoming me to the command offering any assistance with my arrival. I responded that I needed to report in July, because my spouse Calvin was attending the U.S. Army Sergeant Major Academy resident course and would not graduate until May. I was granted permission for the delayed report date.

When I arrived at Human Resource Command, I had my own ideas about how I would approach the assignment. What I didn't know yet was the depth of the opportunities Human Resource Command had in store for me. During my first week, I was scheduled for an office call to meet with the Enlisted Personnel Management Division, Sergeant Major Lynice Thorpe (Noel) in person.

When I arrived at the office suite, it had four large office spaces, decorated with executive furniture, a waiting area, meeting room and full kitchen. Her assistant, a Staff Sergeant, met me at the front desk and informed her of my arrival. Sergeant Major Thorpe (Noel) came out, greeted me, welcomed me to the command and introduced the staff. She explained who worked in the offices, their responsibilities and a little about the workflow of the office. We eventually made to her office. As I entered, there was a wood table with a glass top and four chairs near the main desk, a huge display of her accomplishments, awards, many coins of excellence, framed Command flags from previous assignments, plaques with thank you inscriptions, and a large window with an overview of the main entrance to the building. I thought *this is how they do it here*; my assigned workspace was a small cubicle inside a large room of divided work area spaces.

She welcomed me with professionalism and care, immediately asking if my family and I were settled and if there was anything I needed to feel supported. She had prepared a brief that laid out the Command structure, which was made up of eight Divisions, five of them under Enlisted Personnel Management Division, Operations Management Division, Force Alignment Division, Force Sustainment Directorate, Readiness Division, Enterprise Modernization Division. She encouraged me to set up appointments with each division to gain a deeper understanding of how the organization functioned and how those resources could support our overall mission.

I was initially assigned as the Transition Branch Sergeant Major, but because of my operation experience, she decided to realign me to another position. She stated "Bottom line up front, I'm giving you 30 days to get your feet under you, understand your role and responsibilities and then I expect you to be ready to roll and be prepared to lead and serve. You will be going to units representing the command. You good?"

It was clear from the outset that leadership at Human Resource Command valued a collaborative and informed approach. I immediately went to work. I was assigned to the division responsible for managing the careers and family considerations of all Active Component and Army Reserve Soldiers ranked Sergeant and above more than 374,000 Noncommissioned Officers and enlisted Soldiers around the world.

As a senior leader within this division, I played a critical role in ensuring Soldiers were matched with the right assignments at the right time, balancing readiness, family needs, and career development.

Specifically, I was the Operations Sergeant Major within the Operations Management Division. My role involved overseeing several key personnel programs, including Compassionate Assignments, Exceptional Family Member Program (EFMP) placements, and the Security Centralized Background Screening process. These programs were essential to ensuring readiness and stability for both the Army and the individual Soldier.

The operational tempo was intense, an average of 6,800 enlisted interactions occurred daily, with over 60,000 school seats managed annually and more than 2,500 individual Soldier actions processed, from routine assignments to life-changing compassionate reassignments. Thirty days later, as Sergeant Major Thorpe (Noel) promised, she called me to her office and gave me my first briefing assignment. I assumed since I was new in this assignment, I would start with a small mission and work up, but that was not the idea Sergeant Major Thorpe (Noel) had in mind.

My very first brief was at Command and General Staff College, the Army's senior tactical school at Ft. Leavenworth, Kansas. She explained that every Sergeant Major under her guidance would brief the Sergeants Major at the Pre-Command Preparation course quarterly on the processes and procedures of their individual division. She asked me to see her assistant for my read-ahead book and it would provide me with all the details of travel, lodging, and command information. I scanned the read-ahead book and learned that a second Sergeant Major would be going with me to brief. What a sigh of relief when I read another Sergeant Major was scheduled to attend with me, and he was well versed in briefing this course. We arrived and got our schedule together and proceeded to the school to meet with the afternoon class of students and the school commanding staff.

Briefing was not new to me but briefing as a Human Resource professional was. I had spent countless hours preparing reading regulations, analyzing reports, and cross-referencing personnel actions until they became second nature. This wasn't just another task on my calendar; it was a defining moment, and I knew I had to own it. There's something about standing in front of your peers that makes everything heavier—the expectation, the judgment, the unspoken comparison. And yet, I welcomed it. My colleague briefed me first. I listened intently not just to their words, but to the subtle reactions of the room: a slight head nod, a glance at a watch, a tap of a pen. I watched their posture, the shift in tone, and the questions that followed. All of it gave me what I needed, not just insight, but clarity on how to connect. I silently adjusted my approach. I wouldn't mimic; I would elevate.

When the class leader introduced me, I walked to the front with a calm, steady stroll. Butterflies weren't new; I had felt them before every major moment of my career before stepping into deployment briefings, before leading formations, before confronting difficult conversations. But I had learned to trust the preparation. I reminded myself that nerves didn't mean I wasn't ready. They meant I cared. I opened it with purpose.

My voice steadied as I walked through each point, briefing on human resource policies with the confidence of someone who had lived the field and studied the doctrine. I didn't recite; I *translated*. I made the data relevant, the policy personal, the acronyms understandable. I connected real-world consequences to every decision we made behind a desk. Because for me, human resources were never just about numbers on a roster, it was about lives, careers, families. I could see the shift in the room as I spoke. They were tracking. They were learning. I saw the nods, the pens picking up, and the posture leaning forward. I had their attention not because I was the loudest, but because I was *authentic*. I spoke from a place of experience and care. I wasn't trying to impress, I was trying to inform and empower.

When I finished, there was a pause. Not the awkward kind, but the reflective kind. And then came the questions, not to challenge, but to understand deeper. I welcomed them. I had answers, and when I didn't, I promised to follow up. Because being a professional isn't about knowing everything in the moment. It's about knowing how to find the answer and being accountable enough to follow through. That day I walked away with more than a completed task. I walked away affirmed. I proved to myself that growth isn't about comfort, it's about showing up when the stakes are high, when you're the one expected to deliver, and when you're the one representing your profession, your team, and your own sense of worth. Failure was never an option, not because I was afraid of it, but because I had already decided success was my standard.

Once I returned, Sergeant Major Thorpe (Noel) had already heard about my briefing and wasted no time letting me know she was impressed. With a warm but pointed smile, she said, "Okay, I see you." From that moment forward, doors began to open in ways I hadn't expected. She assigned me to more high-profile briefings, not only as the Operations Sergeant Major but also in her place when her packed schedule filled with the demands of caring for Army personnel prevented her from attending. I was entrusted to represent our command at the Adjutant General Ball in Europe, the Ft. Hood Leadership Conference, the Transportation Command Leadership Workshop, and, to my complete surprise, on the stage at the United States Army Sergeants Major Academy, a place I never imagined I would brief. It was an honor to be a trusted agent of the Command, and more personally, of Sergeant Major Thorpe (Noel) herself.

About two years into that assignment, the invisible hand of God moved once again in my life. I have always done everything for the glory of God, waiting on His gentle spirit to lead and prepare the way for my next journey. In 2019, Sergeant Major Thorpe (Noel) was selected as the first Black woman to serve as the 16th Command Sergeant Major of Human Resource Command, an historic and powerful moment. Her promotion, however, created a six-month leadership gap in the Enlisted Personnel Management Division position that needed to be filled until a new Sergeant Major could be chosen through the nominative process, which involved record reviews, prior command experience, education, and in-person interviews.

Though I didn't expect it, she selected me as the interim Enlisted Personnel Management Division Sergeant Major. That selection had to be approved by the Human Resource Command General Officer, Major General Jason T. Evans, now Lieutenant General (Ret.), and after a week of careful consideration, it was made official. This was monumental. As an Active Guard Reserve Soldier, I was stepping into a nominative position historically reserved for active component Sergeants' Major. The role had never been filled by a reservist until now. But God. I was very hesitant to move into the office. I was going to perform the duties from my current office space, but a mentor of mine, Brigadier General Tawanda Young, reminded me that you lead from the front and you must be seen and heard from the front. She said, "You will move into the office and provide that leadership."

I followed her guidance and moved into the office humbly, and with a sense of purpose. I became the first Active Guard Reserve Soldier to serve as the Human Resource Command Enlisted Personnel Management Division Sergeant Major. I decorated the space with my coins of excellence, military awards, and one special picture: a 24 x 36 image in bold colors of all the Marvel superheroes standing in full costume, each holding the tools of justice and protection. Underneath the frame a single sign in patriotic red, white, and blue. Each corner of the sign was a star, four in total, and in between the stars read simply, "It Takes All of Us." For me, those four stars represented the four angels standing at the corners of the earth, holding back harm, keeping the land and its people safe. That image and message embodied my leadership: unified, protective, and servant-hearted. I can never fully repay the Human Resource Command Team for believing in me and providing a platform I never imagined. It was the most epic, humbling moment of my career.

Today, Lynice and I are bonded, forever sisters by purpose, chosen to carry out a mission greater than ourselves. We often reflect on those sacred moments, knowing we didn't just serve, we showed what was possible. Two Black female Sergeants Major, standing in harmony, always camera ready, serving the Army and its families with excellence, strength, and unshakable grace. That decision to accept God's will turned out to be one of the best of my career. The assignment that gave me pause became the highlight of my Army journey. It challenged me in new ways, expanded my leadership capacity, and connected me with people and missions that truly reignited my passion for service. What began as a difficult moment of uncertainty became a defining chapter of growth, impact, and fulfillment. Sometimes the detour is the destination. And in my case, choosing to embrace the unknown became the catalyst for one of the most rewarding seasons of my military life.

During my tenure at the United States Army Human Resource Command, I saw firsthand how the Army sustains its force. General (Ret.) Lloyd James Austin III, 28th Secretary of Defense, once told me, "Sergeant Major, we all will have to take the walk one day." Those words resonated deeply. Command Sergeant Major Robert Boudnik also advised, "You need to sprint to the end." That meant finishing strong. The Army Song took on a new meaning for me: "And the Army goes rolling along." When you transition out, the Army will continue. The organization thanks you for your service, but the mission must go on, filling the ranks with new Soldiers. Understanding this reality helped me transition with pride, knowing I had given my best while preparing for the next generation to lead. The Army shaped me into a dedicated, disciplined leader. My success came from staying green, remaining compliant, prepared, and committed to excellence. That mindset ensured that I was always ready for the next challenge, and it remains with me even beyond my military career. The work we did had a transformative impact on careers and lives, reinforcing the Army's enduring commitment to people as its most precious asset.

The Army took me places I never dreamed of: Alaska, Egypt, Dubai, Abu Dhabi, Cairo, Uzbekistan, Lebanon, Southwest Asia, South Korea, Germany, Jordan, Bahrain. Each place had a rich history, traditions of military service, and a culture that broadened my understanding of the world. One of the most profound moments of my life was being baptized in the River Jordan, standing where Jesus was baptized, feeling the spiritual weight of history. I placed my feet in the Dead Sea, a moment of peace amidst a life of service. These are the rewards of serving not just the places you see, but the people you meet, the lives you touch, and the sacrifices that remind you that freedom is never free.

My service was recognized with an extensive array of medals, ribbons and devices. This final chapter in my military journey wasn't just a professional milestone; it was a deeply personal one. It reminded me that even in times of crisis, purpose finds a way. Human Resource Command allowed me to retire on a high note, knowing I had poured into the careers and lives of thousands of Soldiers and their families, and that my final contribution helped shape the future force of the United States Army.

Identifying that knowledge gaps hinder professional growth, I addressed the disparity in training between Active and Reserve components. While the Army Reserve exists to support the Active force, Medical Supply Specialists were often excluded from USAMMA's complex Class VIII logistics missions due to the long-term nature of those assignments. To bridge this gap, I collaborated with USAMMA leadership at the Division level to establish the Medical Logistics Overview Program. I managed all funding, logistics, and certification, allowing Reservists to participate in critical fielding and inventory windows. This program, sustained during my tenure at ARMEDCOM, ensured our Soldiers possessed the technical proficiency required to rapidly deploy and sustain decisive power worldwide. I also pushed myself to grow profession-ally. I enrolled in advanced leadership courses, like First Sergeant course, and Battle Staff. I pursued higher education at least once a year and stayed active in the community. I became intentional about preparing myself not just for the next rank, but for life after the uniform.

This assignment concluded on June 1, 2020, right during the global COVID-19 pandemic. A time when the world was shutting down, jobs were scarce, and uncertainty was everywhere. But even in the season of adversity, God opened doors and made a way. I was blessed to transition directly into roles that continued to make a difference, both in service to others, and in alignment with my calling.

"People first. Soldiers always."

The Test

Making friends and facing roadblocks.

CHAPTER 6

WHAT YOU SEE IS NOT ALWAYS WHAT YOU GET

*"When you walk through the fire,
you shall not be burned."*
Isaiah 43:2 (NKJ)

Women of all colors have served in every U.S. War since the American Revolution, often overcoming both racism and sexism to serve. In Revolutionary and Civil Wars, many served as spies, nurses, and support staff often unofficially. Women have been used to change targets on a live firing range for aviators' qualification training.

Harriet Tubman, despite having some of her historical accomplishments erased from the historical website, famously served as a Union spy and scout during the Civil War, and was successful at commanding Soldiers. In a time when women were prohibited from serving in the military, Cathay Williams disguised herself as a man and enlisted in the U.S. Army on November 15, 1866 under the pseudonym "William Cathay." She joined the 38th U.S. Infantry Regiment, one of the units composed of African American Soldiers known as Buffalo Soldiers. During World War I and II Black women served in segregated units. In WWII, the 6888th Central Postal Directory Battalion was an all-Black, all-female unit responsible for clearing a massive backlog of mail in Europe. Major Charity Adams was the first African American woman to be an officer in the Women's Army Auxiliary Corps. It wasn't until President Harry S. Truman signed into law on June 12, 1948 that women were allowed to officially become a part of the military, to receive the same benefits, pay and opportunities as their male counterparts.

Integration of the military in 1948 led to more opportunities, though progress was slow. The cap (the number) for women to join was 2%. Today Black women serve in all branches and ranks, including leadership roles. Black women make up a significant portion of women in the military, especially in the Army and Air Force. We are often overrepresented in enlisted ranks, but underrepresented in officer corps. We still face barriers to promotion, racial/gender bias, and issues like hair regulations and discrimination. Regardless, Black women are trailblazers and have broken the glass ceiling in leadership: Lt. Gen. Nadja West, First Black woman Surgeon General of the U.S. Army; and Col. Merryl Tengesdal, First and only Black woman to fly a U-2 spy plane.

Serving alongside peers, subordinates, and senior leaders in the military wasn't just a test of my skills, it was a test of my patience, endurance, identity, and resolve. As a black woman in uniform, that test came with added layers often invisible, but always heavy. While serving in positions of responsibility and authority, I quickly learned that not everyone was excited for me to be there. The reality is, just because something is working doesn't mean it's working *for you*. Behind the salute and handshake, there were often rooms I wasn't meant to enter, not because I lacked ability, but because I didn't look like the image they had in mind for leadership. In those spaces, white males often chose comfort over competence, cliques over character. Professional development became a privilege for a chosen few, not a right for all. I watched as groups formed in offices, corners, and conference rooms sharing dreams of promotion and power, carving out futures together while leaving me with what was left, never what was deserved.

What You See Is Not Always What You Get

They never considered the impact on my family or the legacy I was trying to build. Instead, they whispered lies, spread misinformation, and smiled in my face while undercutting any opportunity that could expand my reach, my impact, or my voice. I was rarely seen as someone who had *earned* their place. Instead, I was painted as someone who simply got lucky, an outsider who didn't understand how the "real" game was played. They gathered behind closed doors and on conference calls, discussing where I *should* serve but never placing me where my experience and leadership could make the greatest difference. I was never brought in, never shown the ropes, never given the benefit of inclusion. To them, I was "not one of us."

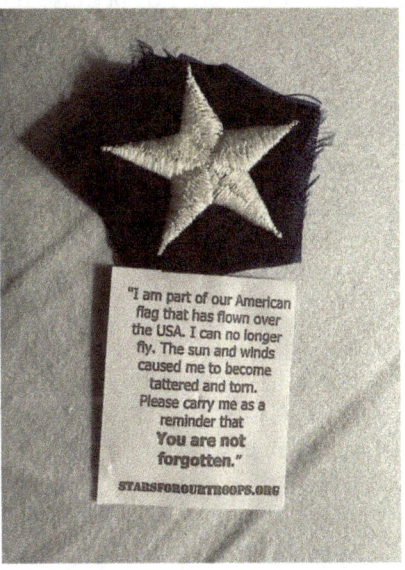

An example of the memento that I gave to many of my mentees. The tag says "I am part of our American flag that has flown over the USA. I can no longer fly. The sun and winds caused me to become tattered and torn. Please carry me as a reminder that you are not forgotten."

The things working against you are invisible, but God, He saw what they didn't. He opened doors they couldn't open. And when I walked through, He shut them behind me so no one could drag me back out. God placed me in rooms where decisions were made in spaces far beyond the reach of my peers. He gave me assignments that couldn't be revoked, missions that no one else could sabotage. He gave me *favor*, and with that favor came the *impact* of the power to change lives, elevate others, and create systems where fairness could breathe.

Because of Him, I was able to help service members, receive what they had earned: recognition, promotion, and fair treatment. I became a vessel of transformation, not because others believed in me, but because *He* did. And every assignment meant for harm; He turned into purpose. Patience was tested when I had to navigate the silence in rooms where my voice should have been heard. I learned to choose my battles wisely, and when to make silence uncomfortable enough to command attention. *Endurance* went beyond physical training. It meant enduring assumptions, being second-guessed, or overlooked then still showing up, still leading, still standing tall. That's a different kind of stamina.

As a Black woman, I often carried the dual burden of proving I belonged and representing everyone who looked like me. The pressure to be excellent wasn't just personal, it was historical. My ability wasn't just measured by how well I performed; it was measured by how I navigated systems not built for people like me and still managed to thrive. Still, I led. Still, I uplifted others in the process. Education gave me power, but I had to learn how to use it wisely in rooms where credentials didn't always translate to credibility. I had to master the balance of humility and assertion, because knowing too much can make some people uncomfortable. I carried the weight of decisions that affected missions, careers, and lives. But I also carried the responsibility of being an example, especially for younger women of color watching silently, wondering if they could one day lead like me. Every test refined me.

Being a Black woman in this space meant knowing I would be questioned, but still showing up with answers, clarity, and compassion. It meant leading with grace under pressure, and with grit under fire. It meant transforming every challenge into a legacy. Not just for me, but for those coming after me. Stand fast, and stand tall. Your purpose will be revealed in the midst of your journey. He will guide you and keep you.

> *"If you have to say you're in charge, you're not in charge."*

The Approach to the Inmost Cave
Getting closer to her goal.

CHAPTER 7

THE PIVOT SPACE

*"In this world you will have trouble. But
take head I have overcome the world."*
John 16:33 (NIV)

I want to share a part of the journey we don't often highlight. The part that happens not on the battlefield, but behind closed doors. Not in combat zones but in conference rooms. Not always in chaos, but in silence. It's what we call in storytelling the "Approach to the Inmost Cave," that moment where everything you are is tested in darkness, alone, without applause. For me, that cave wasn't a place, it was a system. It was in rooms where I had earned my seat but wasn't welcome to speak. Where my experience spoke volumes, but whispers tried to write my story before I could tell it.

There is a moment in every leader's journey when the path narrows, not just because of the mission ahead, but because of the weight of what must be endured to move through it. For me, that moment wasn't a single night, a loud confrontation, or a dramatic turning point. It was a season. A long, quiet walk into rooms where I was the only one who looked like me. Where authority was granted but not always respected. Where silence often screamed louder than words. I had earned my place. My record, my performance, my education, and my faith proved it. But not everyone was excited to see me there.

THE INVISIBLE HAND

Sometimes the hardest terrain isn't the battlefield, it's the boardroom. The briefing. The hallway conversations you're not invited to. I learned fast that although things were functioning, they weren't always working in my favor. I watched white male peers choose color over character, cliques over competence. I witnessed leadership being quietly handed off in informal gatherings, and future roles secured with backslaps and inside jokes. When it came time to extend opportunity, those like me were offered only what remained. Smiles were delivered publicly, but behind closed doors, seeds of doubt were planted to ensure I had to earn every inch twice. I became the target of false narratives, not because I was failing, but because I was excelling. This is the woman the cave did not break. This is how I lead so others can rise. Rumors spread because the truth was a threat. My silence wasn't a weakness, it was strategy. But when the time was right, I made the room uncomfortable with that silence. Just enough to shift the energy. Just enough to remind them, I wasn't going anywhere, and my steps are ordered and no weapon formed against me shall prosper.

Endurance became more than physical, it became mental, emotional, and spiritual. I sat in rooms where my voice should have led, where my ideas could have advanced the mission, but my seat came without a welcome. I was tasked with decisions but excluded from the conversations. Still, I led with precision, purpose, and presence. I showed up, I outperformed, and I led especially when the spotlight wasn't on me. And behind my quiet confidence stood my leadership philosophy, which never wavered. It became my compass. "Always do what is right; never choose the easier wrong over the harder right. Keep all informed; share knowledge to empower the workforce. Know your staff well enough to elevate them to the next level."

As the Senior Noncommissioned Officer in charge, I carried the responsibility of leadership that others could accept, understand, and grow from. I knew early on that no two Soldiers are led the same way, and adaptive leadership situational awareness was essential. In that inmost cave, I held fast to my identity. I am a mentor, a trainer, and a coach. I hold Noncommissioned Officers to higher standards because standards build character.

I don't just say the Noncommissioned Officer Creed, I live it. I am proud, humble, and nurturing. I carry my professionalism like armor. I demand courtesy, military bearing, and performance, not perfection, but persistence. I didn't just endure. I led through it. I insisted that my leadership be felt even when it wasn't always seen. I modeled it on my communication, my presence, and my fairness. I listened actively and spoke with tact. I raised expectations not just for my Soldiers, but for myself. But let's be honest: the pressure wasn't just personal, it was historical.

Every step I took echoed with the legacy of those who came before me. Every room I entered held ghosts of opportunities denied to others. My confidence had to be carefully measured, because too much of it made others uncomfortable. But still, I pressed forward. My knowledge gave me power, but I wielded it with balance. I took up space without apology and I made space for others to rise with me. My leadership extended beyond rank; it became a ministry. "Leadership is about Soldiering. We don't have bad Soldiers; we have Soldiers with concerns that must be addressed. I will provide them with tools. I will train with them. I will trust them. I will help them become their best selves."

As a Christian leader, I carried my faith like a quiet storm. It helped me identify with others, spark new motivation, and keep my focus on what matters: Prevent, Shape, Win. Skills, knowledge, and behaviors. I saw Soldiers not as problems to fix but as people to serve. And when others sought comfort, I embraced the challenge, because I knew my presence alone was a disruption, one that had purpose.

Even though my commitment to fitness wasn't just about strength, it was and still is about mindset. Fitness is an attitude, I trained to standard, not to time. I emphasized healthy choices for food, because food is a tool for a life lived well, mindsets, and people in your circles and readiness for the real-world demands of duty. Because leadership is a 360-degree responsibility mind, body, and spirit. The inmost cave was the place where fear wrestled with faith, where legacy clashed with comfort and where the call to lead collided with the need to survive. I did not emerge unscarred, but I did emerge unshaken. And more than that, committed. Because it was never just about me. Every test refined me for those who would come next. Every challenge shaped the leader I became, a leader who now walks beside others through their own caves. One who leads not just from the front, but with the full weight of experience, vision, and a calling that cannot be shaken. This is my leadership philosophy in motion.

The Creed of a Noncommissioned Officer

No one is more professional than I. I am a noncommissioned officer, a leader of Soldiers. As a noncommissioned officer, I realize that I am a member of a time-honored corps, which is known as "The Backbone of the Army." I am proud of the Corps of noncommissioned officers and will always conduct myself so as to bring credit upon the Corps, the military service, and my country regardless of the situation in which I find myself. I will not use my grade or position to attain pleasure, profit, or personal safety.

Competence is my watchword. My two basic responsibilities will always be uppermost in my mind—accomplishment of my mission and welfare of my Soldiers. I will strive to remain tactically and technically proficient. I am aware of my role as a noncommissioned officer. I will fulfill my responsibilities inherent in that role. All Soldiers are entitled to outstanding leadership; I will provide that leadership. I know my Soldiers and I will always place their needs above my own. I will communicate consistently with my Soldiers and never leave them uninformed. I will be fair and impartial when recommending both rewards and punishment.

Officers of my unit will have maximum time to accomplish their duties; they will not have to accomplish mine. I will earn their respect and confidence as well as that of my Soldiers. I will be loyal to those with whom I serve; seniors, peers and subordinates alike. I will exercise the initiative by taking appropriate action in the absence of orders. I will not compromise my integrity, nor my moral courage. I will not forget, nor will I allow my comrades to forget that we are professionals, noncommissioned officers, Leaders!

My Leadership Philosophy

I would like to share my personal style of leadership philosophy with all Soldiers. To have a cohesive, avid, and high achieving team, it is essential to incorporate a leadership philosophy as a road map to success and develop an interrelated positive reception of my leadership philosophy. Leadership in today's workforce is an essential characteristic to manage effectively.

The philosophy I manage, live, and lead by is simple, precise, and consistent. First, always do what is right, never choose the easier wrong, over the harder right. Next, keep all informed, share knowledge to empower the workforce and finally know your staff well enough to elevate them to the next level. All leadership has an obligation to this 21st century Army to grow and train their Soldiers. As a senior Noncommissioned Officer in charge I have the ultimate responsibility to have a leadership style that all subordinates can accept and understand. Throughout my military career I have come to realize that all personnel are not led the same way under the same style of leadership and may require adaptive situational awareness to be successful.

Characteristics

I am a mentor, a trainer and a coach; I hold Noncommissioned Officers to higher standards. Standards build character. I am proud to be a Noncommissioned officer. I am driven to serve those below and above me, and I strive to achieve excellence through others. I embody the Noncommissioned Officer Creed. It is a lifestyle and for me it is personal. It was designed to measure you. I persist in holding myself to its guidance and continue to reinforce its value into all Noncommissioned Officers. I am holding its true purpose to be knowledgeable, fair and impartial to all Soldiers. I am professional in appearance, and I anticipate all Soldiers will show common courtesy and execute military bearing daily. My demeanor is one of humility, yet gentle and very nurturing. I have a strong sense of identity and possess a positive force of experience and loyalty. I expect my Soldiers to be high speed, energetic, and enthusiastic. I am an active listener when communicating; I use tact and articulate ideas precisely and concisely. As a senior Noncommissioned Officer, I have set my leadership goals extremely high, and I have high expectations of all leaders. I live by the Army Values and will insist on each Soldier doing the same.

Communication

The absence of effective communication will eventually affect the unit's performance. Personal problems can create serious chaos in a work environment. Individual problems are perhaps the most complex. Whether it's poor communication, or interpersonal conflict. To be an effective leader, you must first learn to follow. I believe a leader should learn it, observe it, adapt it, and incorporate change when necessary. Be effective and willing to lead from the front. Know your Soldiers; your responsibilities as a leader and communicate constantly and always keep them informed.

Dedication

Soldiers come from all different backgrounds in this great Army. As a Christian and leader with a strong belief in Christ, I will identify with them, create new, and innovative ways to improve self-motivation, so they can prepare to focus on Rally Point 32, and get back to the basics, prevent, shape, and win, plus create initiatives to focus on the 20/20 vision of operations. I will persist in instilling a mindset oriented towards working towards an objective until it is achieved. I am the type of leader that not only goes the way but shows the way. We as leaders need to let our Soldiers know that we are proud of them, and how important their selfless service is to this 21st century Army. I will provide my Soldiers with the necessary tools to accomplish the mission by training with them, trusting them in combat, and encouraging them to do the best of their abilities.

Fitness

Fitness is an attitude and it prolongs life. I enforce the Army physical fitness training program, and train to standard not to time. My goal is educating everyone on the importance of keeping fit and making healthy choices to enhance performance that meets the physical demands of any combat or duty position.

Noncommissioned Officers are the backbone of the Army. My philosophy embraces characteristics, communication, dedication, and fitness. These keystones combined are my center of gravity for my leadership philosophy. Each is not all-inclusive, but they are essential to have a cohesive, avid and high achieving team. My intention is that each keystone will provide the unit with focus and an understanding of my expectations of you as a Soldier and as a member of this unit.

"Put the devil under your feet."

The Ordeal
The hero's biggest test yet.

CHAPTER 8

TO EVERY PURPOSE THERE IS A TIME AND JUDGMENT

"The Lord is my strength and my shield."
Psalm 28:7 NIV

Every Soldier prepares for trials in the field. But nothing in my military training prepared me for the battle I would face in my own heart, the kind that comes not with bullets or commands, but with silence, expectation, and a decision that could not be made for me. I had served faithfully, rising through the ranks, leading with discipline, and earning the respect that comes from showing up in the hardest places and still performing with excellence. I sacrificed time, comfort, and pieces of myself for the mission. I believed I would continue until I became a mother. Let me take you to a moment that defined my journey. Not once, but twice.

When I learned I was expecting my first child, I thought of my mom and how she carried and cared for five children. I could not move past that moment of the choice before me. I remember when the paperwork was introduced to me by my leadership, the tone and body language led me to feel like they didn't care, I didn't matter, the unborn child didn't matter, the only thing that mattered was the decision. It felt more like an ultimatum. *If you wanted to stay in the Army, why did you get pregnant?* There was no room for conversation, no understanding, and no support. Just a look, one I can't forget, that said, "Well, you chose this." I wasn't seen as a woman who had served faithfully. I was now "the one who got pregnant." It's amazing how quickly people can reduce you to a choice they've never had to make. Something in me shifted. The clarity hit like a shock wave: I could not hand this child off to my mother; this was my decision, not hers. But the system I was in wasn't made for that kind of choice.

When people hear that I left the Army to have a child, they may imagine a clean decision, paperwork signed, uniforms packed, mission closed. But it wasn't clean. It was messy. It was war. Not the kind you fight with weapons or commands, but the kind that creeps in quietly, asking, *"Are you sure? What if you never get back? What if you lose everything?"* The message was clear, you can serve, but not if it means choosing your child. I was faced with the ordeal of choosing between two versions of myself the Soldier who had trained and proven herself, and the mother who knew that presence in a child's first year is a kind of leadership filled with meaning that is sacred and irreplaceable. I battled fear, *how will I provide?* I battled guilt, *am I betraying everything I worked so hard for?* I grieved, because walking away from something you love even for something greater, still hurts. But I also found resolve. The kind of strength you don't get in the gym or on the range. The kind forged in a mother's gut when she knows the right path, even if it leads through loss. So I laid down the career I had built to hold the life I had to bring into the world. I walked into the fire of judgment, of whispers, of what-ifs, and came out with clarity: my child was not a detour; it is now my new mission.

That was my ordeal. Not a moment of death, but of dying to the image others had of me, to the plans I had once mapped out. It was the moment where I sacrificed status for legacy, and where God met me, not with a title, but with peace. I had no guarantees, but I had a mission greater than myself. In that crucible, something new was born in me. A different kind of leader. One who knew that true strength is not proven by staying in the fight, but by knowing when to choose a higher one. Because one day my child would not remember what I sacrificed, but that I made the best choice for the circumstances and chose him over the Army. That will always be the proudest chapter in my journey. After the birth of my child, I was able to go back into the Army Reserve and continue to fulfill my contract obligations.

Seven years later, after being deployed and returned home from duty, I was back in the same position, on duty and had to make the decision to return home or chapter out of the Army, because I was pregnant with my second child. This time, circumstances were different, not because I was pregnant, but because this time my military contract obligation was completed. I knew the cost this time. I had no more time left to serve, I had to reenlist to get another six-year contract, but I was pregnant, and you can't join the armed forces pregnant. I was determined to serve, but I also knew the truth: the mental scars from the first time had never fully faded. I still remembered the sleepless nights, the second-guessing, the moments I felt like I was watching my dreams drift away, and yet, I also remembered the peace that came after. But let me be clear, I didn't give up. I chose to show up for my children, for my calling, and for the life I was entrusted to lead beyond the battlefield.

Once I gave birth to my child, we both had new life. To re-enlist and start again, I had to go all the way back to the beginning, taking the Armed Service Vocational Aptitude Batter (ASVAB) test to re-enter the service, going through a full medical exam and reclassification in a different Military Occupation Specialty and attending a new skilled proficiency. It was not easy at the age of 32. When I overcame that ordeal and endured everything to finally rejoin the Army, it felt like God had made a way back through a closed door. I found my rhythm again. I wore the uniform with pride. I felt whole. Coming back meant facing more than just new regulations or updated uniforms, it meant facing the mental weight of uncertainty: *Do I still belong? Will I measure up?* But I found peace, the kind of peace that only shows up when you've made the hard, right choice and laid your fears at God's feet. It was mental endurance; "Do you really want to start over again?" And God whispered, "This is not a loss. This is legacy."

The truth is that the Army trains you to survive the battlefield. But it doesn't train you to survive grief, regret, or invisible pressure. It doesn't prepare you for the mental weight of knowing that every move you make is being measured, not just by others, but by your own hopes, fears, and future. I learned the discipline of silence, the kind where you carry doubt in your bones and still show up with strength on your face. I learned to war against the voice that told me I wasn't enough, I would never make it, because I'd walked away twice, because I'd interrupted my career for something that couldn't be measured in rank or ribbons. I had to rebuild my confidence not in a promotion, but in purpose. I had to remind myself daily that obedience to God and faithfulness to my family was leadership. Even if no one saw it. And I had to learn how to mourn a version of my military career that might never return, while still trusting that something better, something more eternal was being built. That's the mental ordeal of it all. The sacrifice wasn't just professional. It was emotional. Psychological. Spiritual.

But in that long, quiet fight I became someone new. Not just a mother. Not just a Soldier. But a woman who could make peace with the silence. Who could walk away from what she loved twice and still come back with power. Because true strength isn't just surviving the mission, it's surviving the moments when no one sees the war you're in and still choosing to lead with love, faith, and clarity anyway. To every young Soldier standing where I once stood, holding on to dreams in one hand and holding the weight of motherhood in the other, I want you to hear this clearly: you are not weak for choosing your child. You are not less committed because you prioritized motherhood. You are not less of a leader because your timeline looks different. In fact, you are walking a path that requires more strength, not less. You are choosing the kind of leadership that doesn't always get medals but always builds legacies.

To Every Purpose There is a Time and Judgment

Here is what I learned, what I want you to carry with you: Your worth is not tied to a uniform. The Army can shape you, but it does not define you. Your value is not reduced when you take a pause to care for life, especially the one you brought into the world. The mission will always be there. Your child's first year won't. You will never regret choosing a presence in their life. What you invest in them now becomes the foundation for who they become. Don't let anyone shame you for having a maternal instinct. You can be both fierce and nurturing. You can defend a nation and still protect your home. Those dual roles do not compete, they complement each other.

There is life after the pause. I left twice. And both times, God made a way. He opened doors no man could shut. And He refined me in the waiting season. If He did it for me, He can do it for you. Lean into your faith, not just your fear. Fear will tell you you're falling behind. Faith will remind you that you're being positioned, not sidelined. Sometimes, the greatest promotions come through pain, not process. Surround yourself with truth-tellers. You need people who won't just advise you to "stay in," but who will remind you to follow peace, to protect your mental health, and to remember your why.

Let go of the need to prove. You don't owe anyone an explanation for choosing your child. Let your choices speak through your peace, your presence, and your perseverance. When you come back, come back stronger, but come back whole. Don't apologize for the detour. Walk back in with your head high, knowing that your story is your superpower. One day, your child will look back and know they were never an interruption, they were the reason you became the kind of leader who understands sacrifice, compassion, and courage in ways no training manual could ever teach. You are not alone. You are not behind. And you are not done. Keep walking. Keep trusting. Keep becoming.

"Stay the course."

Reward

Light at the end of the tunnel

CHAPTER 9

MAKING A DIFFERENCE

"Do not withhold good from those to whom it is due, when it is in your power to act."
Proverbs 3:27 N.I.V

This chapter is very significant to me. I want to share the stories that gave me that pause, stories that stirred something deep within, stories that remind us of the shoulders we stand on.

PFC Gladys Crawford

PFC Gladys Crawford stands as a trailblazer in American military history, a woman whose courage and commitment broke barriers and forged a path for future generations. In the heart of Florida, during one of the most defining moments in American history, a young woman listened closely as President Franklin D. Roosevelt urged women to step forward to serve, to support, and to carry the weight of a nation at war. Among those who answered that call was Gladys Crawford. She didn't just enlist, she made history. She became the first African American woman from the state of Florida to volunteer for military service during World War II. PFC Crawford defied the limitations of her time, those imposed by both race and gender. Her decision to serve was not only an act of patriotism but also a powerful statement of dignity and resolve during an era marked by segregation and systemic injustice.

Her service began in March 1943, when she joined the Women's Army Auxiliary Corps (WAAC). Later, this auxiliary unit became a full part of the Army as the Women's Army Corps (WAC) marking a critical milestone for women in uniform. From the beginning, Gladys was determined to serve with honor. She completed basic training at Ft. Devens, Massachusetts, graduating as a cook. She was assigned to the United States Army Air Forces, serving in the AAF 60 WAAC Headquarters Company in Walla Walla, Washington, under the leadership of Captain Elizabeth C. Hampton, a commanding officer she spoke of with deep admiration. Her assignments included posts in Miami, Florida, Wendover, Utah, and Sioux City, Iowa, where she completed Bakers and Cook School. She wore a patch she described as symbolizing "Savior of America" and always remembered President Roosevelt's call for women to take over essential jobs, so that men could be freed to train and fight America's war.

Her final military station was at Ft. Dix, New Jersey, and when her service ended in November 1945, she returned home to West Palm Beach, Florida, where she continues to reside. She often spoke with pride about being part of the transition from WAAC to WAC, no longer an auxiliary but fully integrated into the U.S. Army. Gladys never sought attention for her service. She didn't speak publicly about her experiences for decades, choosing instead a life of service to others in her community and family. She never married or had biological children, but raised her sister's four children after her passing. Among them was her nephew, retired Sergeant First Class Shawn Gibson, whom she considers a son. He has become one of her greatest champions and accepted a long-overdue honor on her behalf.

I came to know about her through her great-great-niece, Alicia Gibson. Alicia and I found ourselves at the same beauty salon many

Making a Difference

times, sharing stories the way people do when they're waiting under dryers or getting their hair styled. One day, I walked in wearing my uniform, and Alicia looked surprised. "I have been talking to you all this time, and I did not know you were in the Army," she said. Then she added, "My great-great-aunt served in the Army too." That opened the door. I asked her when her aunt had served. When she said it was during World War II, I immediately asked whether she had been a part of the Women's Army Corps (WAC) or the 6888th Central Postal Directory Battalion. Alicia wasn't familiar with either, so I told her how significant those groups were especially the 6888th, the only all-Black, all-female battalion deployed overseas during WWII. They saved the morale during the WWII by processing a massive backlog of mail. It was the only unit of its kind sent overseas, led by Major Charity Adams.

Alicia later brought in some pictures and articles about her aunt. The moment I saw them, I knew she deserved to be recognized. Alicia mentioned that her community in Florida celebrated her service every year. I told her that it was wonderful and that I had friends who could help get her story documented and on display at the Women's Military Memorial Museum in Virginia. One friend, still on active duty in Virginia, was able to make that happen. He sent photos and times of the exhibit so Alicia's family could attend and see her legacy honored. But before any of that could move forward, I needed to involve the Office of The Adjutant General, the Army's office responsible for retirees, personnel records, orders, awards, and much more.

Alicia was excited but surprised when I said we'd start first thing the next morning. I had just been selected as the interim Enlisted Personnel Management Division (EPMD) Sergeant Major, a role that placed me over four divisions and eight directorates at the Human Resources Command. I was in the right seat, at the right time, for the right reason. Though I never met her in person, I did get the opportunity to speak with her on the phone. She was 103 years old. I introduced myself and told her I was in the Army. She replied, "I was in the Army too." She told me how she served in the Women's Army Auxiliary Corps. The pride in her voice gave me goosebumps. But then she said something that broke my heart, someone had broken into her house and stolen her awards and her discharge paperwork. When I asked if she meant her DD-214, she said, "Yes, I think that's it." She no longer had it and wanted it back. I told her I worked at the Human Resources Command and that I would ensure she received a reissued DD-214 and her awards. She was overwhelmed with gratitude, repeating "thank you" over and over.

That moment sealed my mission. I contacted Sergeant Major J. Williams and brought it up in our Monday meeting. We reviewed hot topics every week, and this quickly became a priority. I also reached out to a few people who could assist me with moving things forward. One of them was retired and worked alongside me at one point in time of my career. She had once shared her story as the last enlisted WAC to retire and was featured in the documentary *Unsung Heroes*. She is a mentor of mine, Command Sergeant Major (Ret.) Cynthia Pritchett. I asked her about getting Gladys inducted into the Hall of Fame. She told me to submit her story to the committee for review and possible selection. This honor is not easy to achieve, and I trusted God that the committee would see value in her and honor her service. I submitted her story, and a few months later, I received the news: she had been selected. Then came the moment I had hoped and prayed for. A message arrived from Beth Spitzley, with joy and significance. It read: "It is my honor to inform you that PFC Gladys Crawford has been selected to be inducted into the U.S. Army Women's Foundation Hall of Fame. Please see the attached letter for more information and details. A copy of this letter will also be mailed to you next week, but I wanted to notify you of this honor as quickly as possible. Please reply to me so that I will know this information reached you."

The service records that were ordered supported three medals: the Women's Army Corps Service Medal, the American Campaign

Medal, and the World War II Victory Medal. Based on all records mailed to us, no evidence was found that she served in the 6888th but her story stands firmly on its own. The awards team was prepared to order her medals; all that was needed was a mailing address.

At 103 years young, Gladys Crawford continued to inspire all who met her. On her 101st birthday, the City of West Palm Beach presented her with a Key to the City. She remained fiercely independent until she passed July 2021. She kept in touch with the family she once worked for as a chambermaid over 50 years ago, now embraced as one of their own. Her mother, who lived to be 101, passed on a legacy of longevity and quiet resilience that Gladys carried forward.

When you meet a hero, someone you've never met, yet whose legacy shaped your journey you realize that your own service was made possible by theirs. PFC Crawford is that kind of hero. Her quiet courage made room for others to follow, to lead, and to serve with pride. Today, we honor her legacy not just for her service in uniform, but for the strength of character, perseverance, and grace that continue to inspire. Her story is a vital chapter in the broader narrative of American history one that reminds us that true service transcends boundaries and that heroes often wear their valor with quiet humility. She has always lived a life of selflessness, humility, and strength. Public recognition never mattered to her, but her service and sacrifice are now forever etched in history. Her story, once held close and quiet, now speaks volumes for generations of women and Soldiers. Through her courage and conviction, Gladys Crawford became not only a Soldier but a symbol for dignity, service, and the unwavering spirit of those who paved the way so others could serve with pride.

THE INVISIBLE HAND

Private First Class (PFC) Norman Dufresne

In 2010 I was assigned to the 330th Medical Brigade at Ft. Sheridan, Illinois. It was my first duty station as a Sergeant Major, and with that came many responsibilities, including the solemn duty of serving as a Casualty Assistance Officer (CAO) or a Casualty Assistance Notification Officer (CNO). Every unit has what is known as a "red cycle," a period when members are placed on alert in case they are needed to notify a family of a veteran's passing or assist them in navigating in the aftermath of their loss.

Just a week after completing the required training, I received my first assignment as a Casualty Assistance Notification Officer. I initially thought it was a joke when I was called the very Friday I graduated from the course. But it was no joke. My first case was to notify the children of a Sergeant Major in Chicago of their father's passing. It was a heavy responsibility, but I carried it out with the professionalism and compassion it required.

However, just two days later, I was called upon again this time for something much bigger than I could have anticipated. I was tasked with assisting in the notification and identification of Private First Class Norman Dufresne, a Soldier who had been listed as Missing in Action (MIA) since the Korean War. This was a repatriation case, something I had never experienced before.

Initially, there was some hesitation about my involvement beyond the standard three to four days authorized by policy. However, my personnel officer, Major JB Galbert, recognized the significance of the situation and advocated for my continued participation. Command ultimately honored me to take part in what would become one of the most humbling and rewarding experiences of my career. In October 2013, I was formally assigned as the Casualty Assistance Officer to notify the Dufresne family that human remains had been identified as belonging to their loved one. I met with a civilian officer who had worked extensively on the case. He presented a detailed packet that documented how the determination had been made. The packet included maps, blood sample analyses, body part identifications, and notes collected from the family dating back to 1953, when they had first provided information in the hopes of one day finding Norman.

Norman Dufresne had been one of twelve children. By the time of this notification, only one of his siblings was still alive, Mrs. Claire Weber. Mrs. Weber was an 82-year-old veteran nurse who had dedicated more than 70 years of her life to caring for wounded service members and the sick in hospitals from Leominster, Massachusetts to Chicago, Illinois. She was the sole surviving family member who could officially identify Norman. Once Mrs. Weber confirmed that the remains were indeed her brother's, she signed off on the paperwork to bring Norman home for a proper burial at the family site in Leominster. However, she had one request. She wanted me to be her escort for the entirety of the process. If the military would not allow me to stay, she intended to contact the mayors of both Leominster and Chicago to secure my presence. Recognizing the significance of this moment, Command permitted me to remain with the family for the full nine days required to give Norman the homecoming he deserved.

Boston, a deeply patriotic city, went above and beyond to honor Norman's return. Mayor Mazzarella of Leominster ensured that this repatriation became a community-wide event. The Veterans of Foreign Wars (VFW) post in Leominster had 52 white crosses outside their building, commemorating the 52 service members from the area who had been lost in various wars. Private First Class Norman Dufresne was the first veteran among them to be found and brought home. His cross was the only one ever officially removed and presented to his family. The homecoming was an unforgettable sight. Delta Airlines transported Norman's remains from Hawaii, where he had been buried in an unnamed grave in The Punch Bowl National Cemetery. A Sergeant First Class escorted Norman from Hawaii to Leominster, and the

Boston National Guard Honor Guard was assigned the mission of rendering full military honors. The family planned four significant events to honor Norman. First, his body was received at the Boston International Airport from Delta Airlines. This crew handled Norman with precious care and compassion. The captain explained that he had the honor of returning a war veteran home after being Missing in Action for over 63 years. He asked the passengers to stand and render honors to the fallen war hero. The passengers on board stood quietly, watched, waited to exit the plane until Norman was escorted from the plane cargo under carriage. Some saluted and removed their caps, others placed their hands over their hearts. The scene was breathtaking, all along the airfield people were standing and rendering honors.

Once the honor guard team secured Norman into the hearse, the two limousines and American Legion honor guard followed the hearse slowly off the airfield and onto the toll express to Leominster. The drive from Boston International Airport to Leominster was usually a one hour or more, but the mayor had ordered a police escort to accompany the family into town. That day, the ride was about a 45-minute journey and was met with overwhelming support. Every bridge along the route, nearly ten in total, was closed off and lined with firefighters, police officers, and community members waving flags and saluting as Norman was escorted home.

Simard Funeral Home was in charge of the arrangements. The hearse and the family were met on the way by over 150 motorcycle riders to ride with the family to the tip of town and were met with a gigantic flag the size of a small building hanging from the extended ladder of the town's fire truck and hundreds of people waving flags, cheering, and holding signs that said, "Welcome home hero," "We love you," "Thank God," and "Number one hero." Then a wake was held at the funeral home. The third event was the laying of his body at John Tata Auditorium at Leominster City Hall where a National Guard changing of the guard took place every two hours. Finally, the official funeral was held at the family church St. Cecelia's, located not far from where the family grew up. The church is one of the oldest in the area, and its history dates to 1933. St. Cecelia's church was beautiful inside and out. It was a multi-colored earth-tone, brick cathedral standing more than 100 feet tall, with stained glass windows all around the church. Inside are impressive works of architecture and breathtaking art of angels and a statue of the Virgin Mary. Once inside, you feel instantly warm and embraced by the many colorful windows, designs and sculptures.

You could feel the peace in the room as if the Lord was saying *welcome home Norman*. It was filled from front to back with community leaders, Washington D.C. officials, senators, veterans, and classmates of both Norman and Mrs. Weber and the remaining family. The city and surrounding communities came together to pay their respects and offer their gratitude for his service and sacrifice. For me, this experience was deeply personal and transformative. Leominster was not a place where many African Americans had settled, so seeing my face on the front page of the local newspaper as a military escort was a moment of reflection. Despite potential challenges, I was met with warmth and respect, and I never felt any form of resentment or hostility based on my race or gender.

Two years later, Mrs. Weber passed away. Her funeral happened to fall on my birthday. When my general officer asked me how I planned to celebrate, I told him I would be attending the funeral of a dedicated nurse who had spent her life serving veterans despite never having worn the uniform herself. His response was one of bewilderment, but I simply replied, "Everybody is not me." Every year since, I receive an email from a legacy registry established by Mrs. Weber's family, inviting me to leave a remembrance in her honor. I participated without fail, knowing how much her strength and dedication shaped not only this mission but also my own life.

Bringing Private First Class Norman Dufresne home remains the most moving event of my military career. To have had the honor of wearing the same uniform he once wore, to present his family with the awards, flag, and ceremony he was due, and to help provide them with the closure they had waited decades for this was the highest privilege of my service.

Wall of Honor

The third most significant accomplishment of my military career wasn't about rank or medals. It was about remembrance restoring a sense of patriotism and honor to families who had given this nation the ultimate sacrifice: their loved ones. By the time I graduated from the Academy in May 2010, I had already learned that leadership was not about rank, it was about responsibility. Assigned to Ft. Sheridan, Illinois, I stepped into my new role as a Sergeant Major with pride and determination. It was my first duty station in this capacity, and the weight of it was humbling. Just months earlier, a tragedy had shaken the Army to its core. On November 5, 2009, a mass shooting at Ft. Hood, Texas claimed the lives of 13 people. The shooter, Nidal Hasan, a U.S. Army major and psychiatrist, forever altered the lives of families and the shape of our military history. Some of the victims had been assigned to units under my battalion. They had been preparing for deployment, simply going through routine pre-mobilization procedures, when the unthinkable happened. Their loss felt personal.

When I arrived at Ft. Sheridan, I was issued a wire chart that mapped out the command's echelon. As I began my initial visits to higher headquarters, I noticed a solemn display in one of the buildings in Salt Lake City, Utah, a beautifully crafted Wall of Honor. I paused and read the captions, only to realize that some of the Soldiers honored there had once served in brigade units that I now managed. I was unsettled by the distance between that sacred wall and the families still grieving in the Midwest at Ft. Sheridan,

Making a Difference

there was no such tribute. No daily reminder of the lives lost or the courage of those who survived and still drilled at our facilities. A memorial wasn't just about grief; it was about grounding us in purpose. Every service member who walked through our doors should be reminded of the cost of our profession.

That day I made a quiet decision: we would build our own Wall of Honor at Ft. Sheridan. A visible, permanent tribute could bring healing and recognition. I wrote a proposal detailing the vision, justification, and potential cost. I broke down the purpose, and the emotional impact it would have on Soldiers and families alike. Although the command approved it, there were no funds earmarked for such a project. That didn't stop me. I dug into regulations, resource channels, and funding requests until I found the right pathway. I submitted the paperwork, followed up relentlessly, and kept the vision alive. At the same time, I reached out to senior leaders by email and asked for their thoughts on where the wall should be placed, what it should be called, and how best to honor those we'd lost. Their feedback helped shape the final decision. Every email, call, and coordination effort was a labor of love and respect.

Determined to keep continuity, I selected the same vendor who had created the original memorial in Utah, ensuring our tribute would match in dignity and design. I coordinated details while managing my daily duties. Months passed. As my permanent change of station neared, I pushed harder to make sure the memorial would be completed in time. It wasn't enough to have the wall built; it needed to be unveiled with the reverence it deserved. So my team planned a ceremony. We invited command officials, community leaders, Soldiers past and present, family members of the fallen, and survivors of the Ft. Hood tragedy. The atmosphere that day was solemn, but proud. When I stepped up to speak, I looked out at the faces of those who had served beside the fallen, and those who had raised them. I spoke not just of loss, but of legacy. I shared why this Wall of Honor mattered, how it stood not only for those

THE INVISIBLE HAND

we had lost, but for every Soldier still walking those halls. It was a promise: that we would never become numb to their sacrifice, never too busy to pause and remember. I had the design modified so that names could be added in the future. It wasn't just about history; it was about a tradition to never forget a fallen comrade. The wall needed to grow with time, as the Army always does. That wall still stands. We could never return to their sons and daughters, but we could ensure that they would never be forgotten.

As you reflect on the stories of Private First Class Norman Dufresne, Private First-Class Gladys Crawford, and the sacred Wall of Honor, may you feel the quiet presence of something greater, an invisible hand guiding their footsteps and ours. These stories are more than history; they are living testaments to faith, courage, and divine purpose woven into the fabric of our lives. Each name etched into the Wall of Honor is a soul who answered a higher calling, stepping forward with unwavering trust in something beyond themselves. Their sacrifices remind us that service is a sacred act, and that through trials and triumphs, God's grace sustains and strengthens us. May these stories inspire your spirit to rise to lead with compassion, to serve with humility, and to live with intention rooted in faith. When you see a name, know that behind it stands a life touched by divine purpose, a story held in the palm of God's hand. Let us honor these heroes not just by remembering their deeds, but by carrying their light forward through love, through service, and through the steadfast belief that we are all part of a greater plan. In their legacy, may you find courage for your journey, peace in the unknown, and hope in every step.

"Look at the big picture."

Making a Difference

The Road Back

We aren't safe yet.

CHAPTER 10

FOR THE TIME IS AT HAND

*"After you have suffered a little while,
God... will himself restore you and make
you strong, firm and steadfast."*
1 Peter 5:10 N.I.V

The road back was never a straight line. It twisted through places I hadn't expected, through moments I didn't see coming. After a lifetime in uniform, a decision was placed before me that would redefine everything: report to a new duty assignment I hadn't requested or retire within 60 days. It wasn't just a career choice; it was a spiritual one. I sought guidance, prayed, and weighed the weight of that moment. I chose faith over fear, not knowing at the time how that single decision would become the catalyst for a chapter that shaped my legacy in ways I couldn't have written on my own. Service has never just been a role or a duty for me. It's the ink that fills every line of my story. It's the reason the pages turn. Even when I stepped away from active duty, the mission stayed woven into my DNA. I never served for applause; I served for impact. I served people. I believed in Soldiers. I believed in purpose. And I believed that my presence, even without a title or rank, could still vary the atmosphere around me. Each assignment affirmed a deeper truth that I was still needed. I still had something vital to contribute. That was my road back not just to leadership, but to a deeper, more whole version of myself. The version refined by trials, healed by faith, and emboldened by grace.

THE INVISIBLE HAND

Preparing for retirement.

The flame within me had dimmed during years of grief, motherhood, and moments of doubt. But it never went out. And in that office, in that chapter, it reignited. Fiercely. Unapologetically. Being back meant more than wearing a uniform. It meant standing as a mirror for others especially women, mothers, Soldiers who wondered if they could still rise. I became the example I once searched for. I stood not just as an enlisted leader, but as a mother of resilience, a sister of strength, and a vessel of legacy. I wasn't there to impress. I was there to illuminate the path for those behind me. I came to understand leadership in its rawest form. It wasn't about titles or proximity to power. It was about presence, purpose, the ability to carry the weight of representation while still considering others. It was about knowing when to fight and when to be still, about carrying both the silence of being overlooked and the responsibility of being seen.

My final assignment concluded on June 1, 2020, right as the world turned upside down in a global pandemic. As doors closed, God opened new ones. I transitioned out of uniform into civilian roles that still aligned with the heart of service. I left the Army, but I never left the call. Because for me, service has always been a covenant not just to country, but to calling. The same invisible hand that guided me through foxholes and family loss, through deployments and disease, through motherhood and missions, has never stopped writing the chapters of my life. And even now as I pause to reflect, I know this is not the end. It never was. I came back to finish what I started, to leave more than footprints. I came to leave legacy, not in ribbons or awards, but in the lives I touched, the voices I helped find their strength, and the lessons I taught without ever saying a word.

God was not finished with us yet, I had heard the saying, "If you want to make God laugh, make a plan," but I did not truly understand it until February 2020. He placed an unexpected blessing into Calvin's life and mine. What began as a temporary act of help became a calling we felt led to accept. Savannah Jane Smith, 2 years of age did not simply come to live with us, she was welcomed, wanted, and lovingly entrusted to our care. Through her, God reminded me that His plans are established long before we recognize them, and that children are sacred responsibilities, placed in our lives for His purpose. Savannah has brought joy, growth, and renewed meaning into our home. As my military service chapter closed, another chapter of renewed purpose began. I thank God for the child He entrusted to us and pray she grows into a smart, bold, joyful, and confident young woman, and know she was chosen, deeply loved, and divinely placed in our story.

Savannah, age 2.

Me and Savannah at the Macy's Day Parade in 2025.

Resilience isn't the absence of hardship; it's the refusal to be hardened by it. It's waking up again and again, choosing to stand tall and lead anyway. The road back was not paved in certainty, but in sorrow, surrender, and sacred strength. In the five years following my retirement, grief became an unwelcome companion. I lost my mother, my younger sister, my oldest brother, my god sister, one beloved auntie, my only uncle, five cousins, and three dear friends, two from St. Louis including my adoptive mother, Mary Reese of Huntsville, Alabama, who passed away while we were together on vacation in Paris. The sheer weight of it all could have crushed me. And yet, I remained standing not because I was invincible, but because something greater was holding me up. The invisible hand of grace, steady and unseen, never left my side. It held me when I had no strength of my own, whispered life into my spirit when I felt empty, and gently guided me back to purpose when everything felt like too much. That hand didn't just carry me through grief; it led me through it, reminding me that even shattered hearts can still beat with meaning. This is the road not just back, but forward. With gratitude. With grace. With the abiding peace that comes from knowing I was built for this.

"Stop and think about it."

Resurrection

The final hurdle is reached.

CHAPTER 11

RETIRED NOT EXPIRED

*"Weeping may stay for the night, but
rejoicing comes in the morning."*
Psalm 30:5 N.I.V.

The year I retired, 2020, was not just a turning point in my personal journey; it was the same year the world stood still. The COVID-19 pandemic swept across the globe, reaching every country, every city, and every community. In a matter of weeks, hospitals were overwhelmed, schools closed, and families were confined to their homes. Businesses shut their doors indefinitely. Travel stopped. Celebrations were canceled. Life as we knew it paused. Preliminary estimates revealed the staggering toll of the virus: at least 3.4 million people lost their lives worldwide. Each number represented a soul, a family, a future interrupted. Even the United States Army, a force built on unity, tradition, and collective discipline, had to adapt.

In its 250-year history, the Army had never experienced such widespread operational change. The cadence of daily life was altered. Training was limited. Missions were reassessed. Remote work became the norm, and the sound of boots marching in formation gave way to quiet offices and virtual meetings. I had seen many transformations during my 37 years in uniform, but this one was unlike any other. The Army I had given my life to, the one that shaped me, challenged me, and called me to rise was shifting before my eyes.

I had prepared for retirement, yes. But I had not prepared for this kind of farewell. Yet in that unfamiliar silence, I found something deeply familiar: peace. God gave me rest, not just physically, but spiritually. Amid uncertainty, He reminded me of something eternal, that my time had come, and that it was okay to walk away. I wasn't walking into a void. I was walking into purpose. The world was in crisis, and yet I knew one thing without doubt: this too shall pass. He didn't bring me this far in life to leave me now. He didn't sustain me through war zones, loss, sacrifice, and transformation just to abandon me at the threshold of change.

I held on to the truth I'd carried through every chapter of service: *I have never seen the righteous forsaken, nor His seed begging for bread.* My retirement during a global pandemic was not an accident. It was divine timing. The world may have called it a crisis, but in my spirit, I knew it was a transition. My chapter in uniform was closing, but the book was far from over. This was not the end. It was the beginning of something new, something led not by rank or assignment, but by purpose and calling. And even as the world stood still, I kept moving forward in faith.

I can't speak for anyone else, but I know with unwavering certainty that my life has been shaped by more than time and circumstance. There have been extraordinary moments, some painful, some joyful, all transformational that reshaped my thoughts, redefined my actions, and rooted my philosophy of life. The calling, the rewards of service, the challenges, the changes, the difficult choices, and the sacrifices required to sustain a commitment to excellence for more than thirty-seven years did not come without cost. But through it all, I rose. I rose because others rose before me. I stood because others knelt in protest, marched for rights, and broke barriers so I could step through doors once locked. I served on the shoulders of women, especially Black women, who dared to dream bigger than the world told them they could, women whose names may never make history books, but who made it possible for me to write my own. They made it possible for me to serve, to lead, and to finish strongly.

And now after the journey, the battles, the burdens, the blessings, I return with conviction. I know without question: my decision to become a Soldier in the United States Army was no accident. It was divine design. It was purpose wrapped in camouflage. And I will tell the world that it was not easy. It was God who kept me. When I stumbled, He steadied me. When I sat alone in silence, fearful of dying or simply being forgotten, He shielded me. When I made poor choices, He rerouted my path. He placed people, places, and moments of refuge in front of me like steppingstones to survival. He gave me strength when I was depleted, wisdom when I was unsure, and purpose when I felt like I was losing my way. And now, I stand humbled, grateful, and forever changed.

There is a force I cannot see, but I have always felt it. A hand that guided me when my own hands were trembling. Doors opened that I never knocked on. My name is spoken in rooms I have never entered. How else do I explain the favor, the opportunities, the inclusion in some of the most life-altering missions and moments? This is more than coincidence. This is the invisible hand of God, the Holy Spirit whispering, "This is the way. Walk in it." Even in moments when I didn't recognize that guidance, or worse, when I didn't appreciate it, I now see that it was always there.

And now that I have crossed the threshold from the world I once knew into life beyond the uniform, I bring back with me the elixir: not riches, not medals, but perspective, wisdom, peace. I still hear echoes, words from mentors burned into my memory like trail markers along the path home, such as "we all will have to take the walk one day," and "you need to sprint to the end." I carry those words like sacred scripture. They shaped how I approached my transition. No fear. No bitterness. Just understanding.

THE INVISIBLE HAND

We are called to serve, but we are also called to know when it's time to let go. The Army Song took on new meaning. "The Army goes rolling along." That line hit me like a wave. Because it's true. After your final salute, after the band plays your last march, the Army continues. It doesn't pause. But the real question isn't what happens to the Army, it's *Did you make a difference? Did you leave a legacy? Do Soldiers still call you, ask for your advice, check on your well-being? Do they miss your presence? Did your service echo beyond your uniform?* For me, the answer is yes. And more than that, I now know that my life was orchestrated for this purpose.

I was born to serve. But just as there is time to serve, there is also a time to walk away, a time to speak and a time to listen, a time to pour out wisdom, and a time to be filled with new vision, a time to pass the torch, and a time to receive the reward. Do I have regrets? Yes. Would I do it all again? Without hesitation. And for that invisible hand that kept me then and now, I give thanks.

On that day at Ft. Myer, Virginia, as the music played and the final salute was rendered, I didn't feel sadness. I felt the fullness of a complete journey. I felt the elixir of the distilled truth of a life lived in service, and the power to pass that truth on. To the Holy Spirit, the invisible hand that guided me through war zones and waiting rooms, through ceremonies and silent prayers, I say: Thank You. For a time such as this. The military changed my life, not just in the way I walk, speak, or stand but in how I see the world, how I face challenges, and how I carry responsibility. It taught me discipline, yes, but also sacrifice, brotherhood, structure, and pain. It showed me what it means to serve something greater than myself, and what it costs to do so.

I have seen things I'll never forget, lost people I will always remember, grown in ways I never expected, sometimes through pressure, sometimes through pride and struggles. The military didn't just shape my habits, it reshaped my identity. It built resilience into my bones and gave me a perspective that I carry into every chapter of life. Even after the uniform comes off, that part of me remains. Strong. Focused. Forever changed. Life is a journey filled with twists, turns, and unexpected moments that shape who we become. My story is one of resilience, faith, and the pursuit of purpose. From humble beginnings to defining moments that carved my path, I have been carried by something greater than myself. Even when I could not see the road ahead, I now understand that an invisible hand was always guiding my steps. Every joy, every loss, every challenge and victory, none of it was accidental. All of it was preparation.

When I planned my five-year transition out of the military, I thought I had it all figured out, structure, stability, sustainability. But within three years, I became guardian of my two-year-old great niece, Savannah. I lost two siblings. I buried my beloved mother. I was unprepared emotionally, spiritually, for that kind of upheaval. Yet even then, God's hand was guiding me. He didn't stop the storm, but He gave me strength to stand in it. He gave me a mission beyond the uniform.

Much of what I endured could have been different if we'd embraced truth over silence. I learned how damaging it is to avoid hard conversations, especially in times of illness or crisis. We think we're protecting our loved ones by withholding the full truth, but we're leaving them unprepared. God calls us to truth not just for ourselves, but to prepare others. The military gave me many things, but it didn't prepare me for the moment the uniform came off. The structure was gone. The rhythm of missions and expectations vanished. And in that void, I had to learn to trust something more eternal. I was no longer a Soldier on mission. I was a woman of faith walking through unfamiliar territory. And once again, God showed up. He showed up in unexpected calls, in small victories, in quiet reminders that He had never left me.

Transition is not an ending, it's an invitation, a holy invitation to new purpose, to deeper faith, and to trust beyond what we can see. I learned the importance of early preparation not just benefits and paperwork but preparing the heart, preparing for what comes next. As Soldiers, we must think beyond the battlefield. As leaders, we must teach others how to live after service, how to be whole. Transitioning during a pandemic added another layer of fear and instability. But even in that, I found peace, not because the road was smooth, but because God had already gone before me.

Calvin, me and my sons.

Throughout my life, I tried to hold everything together. I wore the armor of protector. My children saw it. I admired it, but it taught them to carry burdens too heavy for their youth. I learned that I am not called to carry everything. I am called to trust. That's what I teach them now. It is okay to rest. It is okay to surrender. It is okay to rely on grace. At the heart of my journey is this belief: we are here to serve. We are here to help others find hope, healing, and purpose. Kindness and compassion are not extras; they are how God moves through us.

This is not the end of my story. It's simply another step in the divine path before me. Life will keep evolving. And so will my faith. I don't know what tomorrow holds, but I know who holds tomorrow. I move forward not just as a war veteran, but as a servant, guided by grace, walking faithfully into every assignment my Creator has prepared. The road back was not paved in certainty, but in sorrow, surrender, and sacred strength.

"Don't lose yourself."

Return

The hero heads home, triumphant.

CHAPTER 12

MOVING FORWARD, I AM ALIVE FOREVER MORE

*"He restores my soul. He leads me in paths
of righteousness for His name's sake."*
Psalm 23: 3 N.I.V.

The battlefield had changed, no longer lined with tanks or terrain maps. No sandbags, no convoy manifests, no weapons to clean or rucksacks to carry. Since retiring, my life has looked nothing like I imagined. Instead of predictable peace, it became a winding road of searching, serving, adjusting, and rising again. I became many things, molded not only by experience but by grace and divine purpose: a Soldier for Life, an entrepreneur, an author, a wife, a mother, an Oma (German for Grandmother), a legal guardian, a woman of faith. Through it all, I kept showing up. I kept saying "yes" to the call even when I didn't know where it would lead.

After retirement, I was recommended by the Cadet Command JROTC team at Ft. Knox, to meet with Colonel Samuel Taylor (Ret.), a Senior JROTC Officer in Tampa, who was looking for women to join his team of instructors. The idea of shaping young minds intrigued me. I prepared myself for the challenge, knowing that interviews with multiple principals wouldn't be easy. After several rounds, I was selected as the Senior NCO JROTC Instructor at Sumner High School in Riverview, Florida. The school was brand new, and so was the JROTC program. That meant starting from scratch with limited funding, minimal structure, and an uphill battle to build a meaningful military culture in the classroom. I spent my own money to bring authenticity to the environment, down to the smallest detail. Then COVID-19 hit. Everything turned upside down. Desks were spaced six feet apart. Masks were worn daily. Hallways were marked. Sanitizer stood like sentinels at every door. Teachers became tech gurus overnight, turning their classrooms into virtual command centers. Students, many too young to process the chaos, were separated from peers and screens became their lifeline. Parents became part-time educators while trying to manage jobs and uncertainty.

My new war zone was a classroom. My formation was a room full of kids trying to make sense of a world in crisis. I traded my combat boots for dress shoes, and my operations manuals for lesson plans. But I was still on mission. The halls of Sumner High became sacred to me. They weren't orderly or quiet, but they were filled with potential. I stood in that classroom not just as a retired Soldier, but as a living testament of what resilience and faith can do. I was the proof that someone like them could rise. As an instructor, I wasn't just teaching about ranks or regulations, I was teaching life. My cadets came from all walks of life. Some were filled with hope. Others carried burdens too heavy for their age. To them, I was a rarity, someone who looked like them and had led troops, briefed generals, and survived war zones. I showed them how to speak with purpose, walk with pride, and recognize their worth. This wasn't just education. It was legacy work. Our school even became a pilot for sixth-grade JROTC electives, a rare privilege.

But just six months into my role, I caught COVID-19. And everything changed again. It wasn't just an illness; it was a storm. It swept in with fear, isolation, and the haunting uncertainty of whether I'd survive. Despite decades of military service in dangerous places, despite everything I'd endured, this virus felt like it might take me out. I had chronic conditions, and the medical consensus at that time was grim. I remember driving around, trying to find the courage to tell my husband Calvin. But God brought me through. Though I recovered, I continued to test positive for weeks. I couldn't return to the classroom without risking the safety of others. With a heavy heart, I resigned. I handed over my keys and my computer and walked away from something I loved. Gratitude was there, but so was grief for what was lost, for what never came to be. But through that grief came a deeper clarity. Healing wasn't just about the body. It was about the spirit, the heart, the unseen places that ache in silence.

A week later, while still recovering, I received a call from a contracting agency I had once applied to. They offered me a position supporting U.S. Special Operations Command at MacDill Air Force Base. I accepted, stepping into a role focused on human resources and talent alignment. At first, it was fulfilling. We improved systems, streamlined data, and built processes. But as time passed, confusion set in, communication broke down, and I felt increasingly invisible. The mission lost clarity, and so did I. I was being tolerated, not valued. I resigned again. But when God has a plan, He always opens the next door.

A few months later, I was introduced to AC Disaster Consulting, a woman-owned, majority woman-led firm focused on helping communities recover from disasters. It aligned with my heart. I joined their mobile COVID-19 vaccination team in Hilton Head, South Carolina. We served underserved populations, lived out of hotels, and worked five weeks on, one week off. That work restored my purpose. But life, as it often does, intervened again. My oldest brother Michael grew critically ill while caring for our mother. I returned home to East St. Louis to help and then he passed. The grief was unbearable. My sister-friend Quana reminded me that grief is an emotion with nowhere to go. I had to find space for that emotion, even while caring for my mother and navigating the maze of Medicare, Medicaid, and elder care. It was harder than any operation I had ever led. There is no field manual for watching your loved one's decline.

When the vaccination mission ended, I stayed home, focused on caregiving, and waited for my spirit to rise again. In time, it did. I began volunteering at the Women's Military Memorial in Arlington. I led tours, hosted events, and told the stories of women who had worn the uniform with courage and dignity. It was healing work. It inspired me so deeply that I launched my own business, "A Time and Place," to honor women's service. I designed a T-shirt that declared women have the right to serve, and it resonated far and wide. I also helped my son with our family's cookie business, Beej's Cookies, the best cookie you will ever eat. Life felt meaningful again.

During one of my tours, I met a contractor from the Logistical Management Institute (LMI). Curious, I reached out to a trusted colleague, COL Shawn Edwards (Ret.), who helped me navigate the world of contracting. Soon after, I interviewed and was hired as an executive business analyst at the Pentagon, supporting the Office of the Under Secretary of Defense for Acquisition and Sustainment. I found a rhythm, a team, and a purpose that felt aligned with my gifts and experience. These days, I travel and savor the life I have built. I ride my black and envy-green three-wheeled Slingshot with a supercharged engine, my Concrete Warrior feeling the wind on my face and the presence of God in every mile. I know that I have been shown favor. I have walked through fire and floods and somehow, I am still standing. People ask if I miss the Army. But I never left the mission. I just changed assignments. I still rise early. I still give orders, write reports, plan missions, and mentor the next generation. But now, the impact is different. It echoes in the confidence of a young woman discovering her voice, in the quiet strength of a student who once felt invisible, in the grateful smile of a parent who sees their child thriving under structured love.

That is legacy. Legacy isn't one moment. It's not a promotion, a retirement, or a final salute. It's the decision to keep showing up after the lights are dim. It's continuing to give, even when no one is watching. And while some days I feel the weight of the world, I remember I was made for this. Built by missions and mercy. Refined in both war zones and waiting rooms. My story was never meant to end with a ceremony. It was always meant to evolve. Every salute I earned, every hardship I endured, every lesson I carried it all led here. And I'm still moving forward.

"Put you on your calendar."

A FINAL NOTE

Throughout my life, from childhood through the final salute of my military career, the invisible hand of God has been ever-present, guiding each step, each assignment, each decision. His hand was there when I was a young girl learning the values of hard work and faith, and it was there when I stood at promotion ceremonies, deployment briefings, and in quiet moments of doubt and triumph.

During my tenure at the United States Army Human Resources Command, I witnessed the full power of what it means to sustain a force and invest in people. I embraced that fully. My decision to join the Army took me across the globe from the icy expanse of Alaska to the holy waters of the River Jordan, where I was baptized, and experienced the stillness of the Dead Sea, where I felt God's peace during my service. I earned many medals, ribbons, and citations, culminating in the Legion of Merit. Not for the accolades themselves, but for what they represent: sacrifice, leadership, and grace under pressure.

My body and mind are battered, broken and filled with chronic pain, but still I rise. When I retired on June 1, 2020, as the world stood still during the COVID-19 pandemic, I transitioned not in fear, but in faith. God had already gone before me; my future was already written. That invisible hand, which had moved mountains behind the scenes, opened doors that shut behind me and placed my feet on solid ground even when my world was turned upside down, now gently ushered me into a new purpose. The Army may go rolling along, but I walked away full having given my all, prepared for the next generation, and honored the call placed on my life.

People have asked, "Do you have regrets?" Yes, I regret that I didn't have a mentor early in my career, someone to help me navigate tough decisions and offer perspective when I needed it most. I regret that I didn't give my children the same home base stability as children that I'd had. I regret the poor choices I made in relationships that may have caused mental stress to my children, but I did the best I could as a mother and a provider every day of their life. I made some hard choices without wise counsel, often learning through trial and error. But I wouldn't change any of my experiences. Each one shaped who I am, taught me resilience, and deepened my faith. What I lacked then, God provided in other ways through growth, grace, and the opportunity to now be the mentor I once needed. Every mission, every Soldier I touched, every sacrifice, and every success was part of a divine assignment.

I didn't just serve; I was sent. And through it all, the hand of God was never absent. Never lose your faith; he will come through for you. Don't give in to what people say, find your own way. Trust yourself and be willing to stand for something far greater than yourself. You are worth it.

ACKNOWLEDGMENTS

When I tell people about my childhood, they think it's a fantasy because people of color did not typically have the opportunities my mom and dad provided for our family in the 1950's and 1960's. My parents were always very present in my life, offering sound advice, supporting me through school activities, sporting events, life choices, and playing a crucial role in all my achievements. Without their unwavering support, I never could have accomplished so many things in my childhood and adulthood.

All my love, appreciation, and gratitude to my husband Calvin, and my two sons Bryant, and Stoney, who stood strong with me as I advanced through my military career. A fun fact about my children, both were born in Shiloh, Illinois, at Scott Air Force Base, seven years apart in the same room at the base hospital. I want to thank them for being patient with my military assignments, and for enduring the frequent relocations from state to state. I'm proud of all their accomplishments and I thank God they are both well grounded, happy and respectful young men.

To my Dear Army Reserve Medical Command Military family, United States Central Command family, Leominster Massachusetts family, church families, Soldiers and their families, all my sisters for life and mentees: your presence in my life has been a quiet strength, a steady light, and a divine reminder that God always sends the right people at the right time.

THE INVISIBLE HAND

To all the significant mentors in my life that gave me a shoulder to stand and lean on, I live by these little morsels, quotes and tips in conversation, and mentorship moments you all poured into me. Your guidance came in many forms through lessons, profound praises, coaching, and words of encouragement such as:

- "Sprint to the end."
- "If you have to say you're in charge, you're not in charge."
- "Look at the big picture."
- "Understand the stroke, stroke, poke method."
- "Make a difference where ever you go."
- "Be aggressive."
- "Stay the course."
- Stay green."

Other times, it was spiritual, like:

- "The invisible hand is always working."
- "Let Jesus take the wheel."
- "Put the devil under your feet."
- "Look to the hills which cometh your help."
- "Say thank you when your feet hit the ground every day."
- "God is your strength and refuge a present help in time of trouble."
- "God can put his finger in there and mix it all up."
- "Make change happen right now."
- "Peace."

And often, it was practical wisdom:

- "Be present."
- "Big brother is always watching."
- "Stay camera ready."
- "You got this."
- "What makes you happy."
- "There is no place like home."

- "Three friends to the end."
- "Stop and think about it."
- "You can't sing."
- "Don't lose yourself."
- "Put you on your calendar."
- "I love you this much."
- "The three C's of life: (challenges, changes, choices."

To my mentors: thank you for pouring into me and taking the time to know and share moments that helped shape my life and thank you for always being there when I called or just needed a moment. You are my true inspiration, and I truly thank you from the bottom of my heart.

Special thanks to Jimmie and Deborah Johnson, Jamey Jacobs, Leon Mangum, Colonel Janet Ross (Ret.), Brenda Myers, Quana Wright, Terrence and Lynice Noel and Sergeant Major Melinda Pressley (Ret.) one of the trusted voices in my circle. You've all walked beside me in support, spirit and in truth, reminding me that my story matters not just for me, but for others who need to know that healing, hope, and purpose are still possible.

ABOUT THE AUTHOR

Raised in East St. Louis, Illinois, just east of the Mississippi River, I learned early to dream beyond the boundaries of my surroundings. Born one of five children to Bennie and Lucille Johnson, my foundation was shaped by faith and perseverance values that taught me to move forward even when the destination was unclear.

In December 1982, I answered a call greater than myself and entered the United States Army, committing to a life of service that would define my character for 37 years until my retirement on June 1, 2020. Alongside my husband, Calvin, who served 28 years and also retired as a Sergeant Major, we raised two sons, Stoney and Bryant. We are proud grandparents to my first grandson Bryson and legal guardians to my great-niece, Savannah.

Today, we continue to serve as Department of Defense professionals. I am currently employed as a senior research analyst with the Logistics Management Institute, working under the Secretary of War at the Pentagon.

My life reflects the Hero's Journey, a path of growth that requires faith before clarity and courage before comfort. Through that journey, I learned leadership is not defined by rank, but by responsibility, grounded in integrity, humility, and putting people first. A Soldier for Life and a servant in the Army of our Lord and Savior, this memoir invites readers to trust their calling, embrace life's challenges, and move forward with purpose and conviction.

www.ingramcontent.com/pod-product-compliance
Lightning Source LLC
LaVergne TN
LVHW020138080526
838202LV00048B/3965